Unit Oriented Enterprise Architecture

Constructing Large Sociotechnical Systems in the Age of AI

Andre Milchman

Apress®

Unit Oriented Enterprise Architecture: Constructing Large Sociotechnical Systems in the Age of AI

Andre Milchman
York Center, ON, Canada

ISBN-13 (pbk): 979-8-8688-1144-9 ISBN-13 (electronic): 979-8-8688-1145-6
https://doi.org/10.1007/979-8-8688-1145-6

Copyright © 2025 by Andre Milchman

This work is subject to copyright. All rights are reserved by the Publisher, whether the whole or part of the material is concerned, specifically the rights of translation, reprinting, reuse of illustrations, recitation, broadcasting, reproduction on microfilms or in any other physical way, and transmission or information storage and retrieval, electronic adaptation, computer software, or by similar or dissimilar methodology now known or hereafter developed.

Trademarked names, logos, and images may appear in this book. Rather than use a trademark symbol with every occurrence of a trademarked name, logo, or image we use the names, logos, and images only in an editorial fashion and to the benefit of the trademark owner, with no intention of infringement of the trademark.

The use in this publication of trade names, trademarks, service marks, and similar terms, even if they are not identified as such, is not to be taken as an expression of opinion as to whether or not they are subject to proprietary rights.

While the advice and information in this book are believed to be true and accurate at the date of publication, neither the authors nor the editors nor the publisher can accept any legal responsibility for any errors or omissions that may be made. The publisher makes no warranty, express or implied, with respect to the material contained herein.

> Managing Director, Apress Media LLC: Welmoed Spahr
> Acquisitions Editor: Aditee Mirashi
> Development Editor: James Markham
> Coordinating Editor: Kripa Joseph

Cover designed by eStudioCalamar

Distributed to the book trade worldwide by Apress Media, LLC, 1 New York Plaza, New York, NY 10004, U.S.A. Phone 1-800-SPRINGER, fax (201) 348-4505, e-mail orders-ny@springer-sbm.com, or visit www.springeronline.com. Apress Media, LLC is a California LLC and the sole member (owner) is Springer Science + Business Media Finance Inc (SSBM Finance Inc). SSBM Finance Inc is a **Delaware** corporation.

For information on translations, please e-mail booktranslations@springernature.com; for reprint, paperback, or audio rights, please e-mail bookpermissions@springernature.com.

Apress titles may be purchased in bulk for academic, corporate, or promotional use. eBook versions and licenses are also available for most titles. For more information, reference our Print and eBook Bulk Sales web page at http://www.apress.com/bulk-sales.

Any source code or other supplementary material referenced by the author in this book is available to readers on GitHub (https://github.com/Apress). For more detailed information, please visit https://www.apress.com/gp/services/source-code.

If disposing of this product, please recycle the paper

To Daniel, Kaylee, Nathan, and Sean.

Table of Contents

About the Author xv

Preface xvii

Introduction xxiii

Chapter 1: Problem 1
    Why Everything Is Right and Everything Is Wrong (Simultaneously) 3
    The Grand Ballet of Stakeholders: A Problem That Can Paralyze 7
    The Great Enterprise Mystery: Who Knows What It Looks Like? 13
        Complex yet Unperceived Architecture: The Lack of Clear Vision 17
        The Abstract Nature of Enterprises 19
        The Evolving Life Cycle of Digital Environments 21
        The Need for a Cohesive Strategy 23
    A Love Triangle: Physical, Digital, and Social, All Wanting Attention 24
        The Interplay of Architectures 25
    The Information Masquerade: Does It Really Flow? 28
    Command and Control: Still Around After All These Years 30
        Bridging the Divide: From Direct to Indirect Coordination 33
        Email: An Unintended Tool for Stigmergic Coordination 34
    To Transform or Not to Transform: The Urgency of Choice 36
    Key Takeaways 37

TABLE OF CONTENTS

Chapter 2: Idea 41

Foundations for Change: The Imperative for a New Enterprise Architecture Paradigm..42

Why Systems 45

    Defining a "System" in This Framework..45

    Systems Thinking in Enterprise Architecture......................................47

    Essential Characteristics of Systems 48

    Integrated Systems and Modular Systems...51

    Alternatives to Systems..52

Why Sociotechnical...53

    The Three Foundational Capabilities 56

    Integration of Capabilities in Sociotechnical Systems........................63

Why Large 66

    Scalability of Sociotechnical Systems...66

    Prototype for Large Systems: The Global Enterprise 67

Why Architecture...69

    Focusing on Fundamental Architectural Elements 69

    Clarity and Comprehensibility in Enterprise Architecture...................71

    Integrating Without Obscuring..71

    Promoting a Holistic View of the Enterprise 72

Why Unit Orientation 72

    Beyond Components and Modules 73

    Principles of Unit Orientation..75

Envisioning a New Architectural Paradigm 82

Key Takeaways...84

TABLE OF CONTENTS

Chapter 3: Theory 87

Triadic Evolution: Synergizing Biological, Organizational, and
Technological Forces 89
    Importance in Human Development 90
    Components of Triadic Evolution 90
    Principles of Triadic Evolution...94

Sociotechnical Systems: Interweaving Organization and Technology for
Enhanced Systems Thinking 97
    Brief History of Sociotechnical Systems...102
    Importance and Impact of Sociotechnical Systems 104

Evolutionary Units: Building Blocks of Adaptive and Resilient Systems 106
    Structural Integrity and Control Mechanisms Within Evolutionary Units 109
    Development Stages...110
    Impact on Civilization 113

Coordination: Synchronizing Complex Interactions...115
    The Traditional and Modern Coordination Dynamics 116
    Potential Impacts on Sociotechnical Systems..118
    Types of Coordination in Sociotechnical Systems 118
    Underlying Mechanisms of Coordination: Communication and
    Connectivity..121

Key Takeaways..126

Chapter 4: Model 131

Representations: From Mimicry to Modernity..131

Models: Beyond Blueprints to Constructing Realities 134

Metaphors: Fitting the Most Meaning in the Least Space...............................138

The Journey: Designing the Quest for the Perfect Metaphor 144

The Moment of Truth: Discovering the Metaphor for LSS 147

TABLE OF CONTENTS

The Digital Enterprise Is a City 150
The Digital Enterprise Is a Hospital 153
The Digital Enterprise Is a University 155
The Digital Enterprise Is a Factory 157
The Digital Enterprise Is an Army 160
The Digital Enterprise Is a Fleet 162
The Final Verdict: The Digital Enterprise Is a Fleet 165
Key Takeaways 165

Chapter 5: Architecture 169

Architectural Approach 170
Component Selection Patterns of LSS Architecture 175
    Homogeneous Selection 175
    Heterogeneous Selection 176
    Hybrid Selection 177
    Favoring Uniformity in High-Level Selection Patterns 178
Structural Patterns of LSS Architecture 180
    Bus 181
    Ring 181
    Star 182
    Tree 182
    Mesh 182
    Hybrid Patterns 183
    Contextualizing the Choice 184
    The Case for Extended Star 184
Selecting a Prototype for LSS Architecting 186
Relational Patterns of LSS Architecture 189
    Exchange Relations 189
    Organizational Relations 189

TABLE OF CONTENTS

    Collaborative Relations 190
    Cooperative Relations 190
    Compliance Relations 190
    Investment Relations 190
    Employment Relations 191

Compositional Patterns of LSS Architecture 192
    Distributed Composition 194
    Partitioned Composition 195
    Interlocked Composition 196
    Fused Composition 197
    Compositional Pattern Selection in LSS 199

Relating LSS to Maritime Formations 199
    Armada As a Metaphor for Global Enterprises 200
    Regional Fleets Within the Armada 202
    Ships Within the Armada 205
    Architectural Implications 208

Novel Architecture Style: Unit Oriented Enterprise Architecture 209
    Embodying Business-Technology Alignment 210
    Ensuring Clear Ownership and Control 210
    Enhancing Clarity and Comprehensibility 211
    Facilitating Changeability: Scalability, Transformability, and Flexibility 212

Unit Oriented Enterprise Architecture: Key Definitions 213
    System 213
    Social System (or Collective) 213
    Sociotechnical System 214
    Large Sociotechnical System (LSS) 214
    Crew 215
    Craft 215

TABLE OF CONTENTS

Unit 216
Agent 217
Unit Oriented Enterprise Architecture: Core Principles 217
Principle 1: Harmonious Integration of Realms 218
Principle 2: Digital Within Social Legitimacy 218
Principle 3: Sociotechnical Focus on Social and Digital Realms 218
Principle 4: Purpose-Driven Sociotechnical Wholeness 219
Principle 5: Coherence in Purpose Implementation 219
Principle 6: Flat Network Organization with System-of-Systems Relationships 220
Architecting the Enterprise: Achieving Global Reach 221
Global Reach 221
Selecting LSS Components 223
Composing the Large Sociotechnical System 225
Structuring the LSS 229
Relating LSS Components 231
Benefits of Unit Oriented Enterprise Architecture 234
A Blueprint for a Responsible and Accountable Enterprise 234
Empowerment of Humans via Control over Digital Resources 235
Demystification of Enterprise Architecture 235
Transformability, Scalability, and Flexibility of the LSS 236
Coordination at the Forefront 237
Uniformity of Unit Architecture 238
Optimized Communication Networks 238
Case Studies and Applications 239
Case Study 1: Healthcare Delivery in Smart Cities 239
Case Study 2: Financial Services Network 241
Case Study 3: International Logistics and Shipping 241

Case Study 4: Smart Manufacturing Complex 242
Case Study 5: Universal Application of UOEA Across Enterprise Types 243
Benefits 244
Integrating Unit Oriented Enterprise Architecture into LSS Strategy 246
Implementation Imperatives 247
Key Takeaways 247

Chapter 6: Design 251

Designing Units: Balancing Purpose-Driven Logic and
Structural Uniformity 252
Structural Uniformity 254
Simplifying the Building Blocks: Applications and Suites 254
Backing Services As Extensions 255
The Craft As a Partitioned Composition: Hull and Operational
Applications 256
Integration of Hull and Operational Applications 256
Digital Chain of Purpose-Driven Constructs 257
Purpose Components and Identity Manifestation 259
Function Constructs As Interaction Facades 259
Process Constructs: The Digital Conveyors 260
Structure Constructs: The Crew Framework 260
Order Components and Regulatory Manifestation 260
Culture Constructs: The Repository of Values 261
Memory: The Data Chronicle 261
Storage Constructs: Digital Warehousing 261
Purpose Driven Design: Core Principles 262
Principle 1: Purpose Driven Sociotechnical Wholeness 262
Principle 2: Coherence in Purpose Implementation 263
Principle 3: Uniformity and Accountability in Sociotechnical Units 264

TABLE OF CONTENTS

    Principle 4: Autonomy and Control Within Sociotechnical Units 264

    Principle 5: Composition of Sociotechnical Units 265

    Principle 6: Purpose-Driven Digital Craft 265

    Principle 7: Modular Composition of Digital Crafts 266

    Key Takeaways 267

Chapter 7: Implementation 269

    Translating UOEA Principles into Implementation Practices 272

        Organization: Distributed and Partitioned Composition 272

        Engineering: Development and Integration of Applications 274

    Organizing LSS and Units 277

        Geographical Organization 277

        Multicloud Approaches in LSS Organization 280

    Engineering Units 286

        Uniform Structure of Units 287

        Applications As Rooms in a House 287

        Suites of Applications As Apartments 288

        Vendor-Provided Suites and Applications 289

    Key Takeaways 289

Chapter 8: Interactions 293

    System-Level Interactions 297

        Coordinating Transactional Interactions 298

        Coordinating Organizational Interactions 308

        Coordinating Collaborative Interactions 312

        Coordinating Compliance Interactions 318

        Coordinating Competitive Interactions 320

        Coordinating Cooperative Interactions 321

What Can Be Shared in Cooperative Interactions 323

Mechanisms for Cooperative Interactions 324

Unit-Level Interactions 326

Direct Coordination 326

Indirect Coordination 329

Fundamental Principles of LSS Interactions 334

Principle 1: Coordination As the Essence of Interactions 334

Principle 2: Multifaceted Relationships Through Coordination 335

Principle 3: Cooperative Interactions to Enhance Relationships and Innovation 336

Principle 4: Structured Communication Protocols 336

Principle 5: Indirect Coordination and Communication 337

Principle 6: Asynchronous Communication and Direct Information Access 338

Key Takeaways 339

Chapter 9: Conclusion 341

Rearchitect to Mitigate Existential Risks 343

What's Next 345

Appendix A: Evolveburgh—A City of Innovation and Tradition 347

Appendix B: Bibliography 383

Index 387

About the Author

Andre Milchman, a hands-on enterprise and software architect, boasts a diverse professional background. His journey began in civil engineering, where he honed his skills as a construction engineer in the dynamic oil and gas industry. Like many of his peers in the 1990s, Andre transitioned into the burgeoning field of computer science, earning a degree that paved the way for his certification as a Sun Certified Enterprise Architect. In Canada, Andre has made significant contributions as a software professional and technology researcher. His passion lies in systems thinking, sociotechnical systems architecture, advanced analytics, and artificial intelligence, which is evident in his written works, such as *Enterprise Architecture Reimagined: A Concise Guide to Constructing an Artificially Intelligent Enterprise, The Triadic Evolution: Reimagine Tomorrow's Sociotechnical World,* and *AI Safety and Security: Architectural Context, Perspectives, and Insights.* Andre's collaborative spirit shines through in his coauthored works with Noah Fang, *Prescriptive Analytics: A Short Introduction to Counterintuitive Intelligence* and *The Transformable Enterprise: Enterprise Architecture in a New Key.* Today, Andre resides in Toronto, Canada, where he continues to explore and innovate in his field.

Preface

The author thanks a fictional government agency for its phantom funding, a university that graciously granted a sabbatical despite no prior affiliation, and a reviewer on Amazon for coining a new term of criticism. Gratitude extends to the City Hall of Evolveburgh, the mayor of the city, and some secretive intellectual society.

The journey to writing this book began with a humble idea—unit oriented architecture—shared on my website, method.org. Like a small acorn growing into a mighty oak, that idea sprouted into my first book, *Enterprise Architecture Reimagined: A Concise Guide to Constructing an Artificially Intelligent Enterprise*. This debut attempted to mend the flaws of previous methods, much like duct tape on a leaky spaceship.

The catalyst for the second book, *Prescriptive Analytics: A Short Introduction to Counterintuitive Intelligence*, cowritten with Noah Fang, was a delightfully scathing one-star review on Amazon by Richard Foster. His review, titled "nonsensense," stated, "This book has words, sentences, and diagrams. There is no value for the material beyond that." I was truly inspired by his invention of the new word "nonsensense" to describe my work. This critique propelled us to unravel the mysteries of designing a prescriptive analytics system. We modeled situations, conjured actionable alternatives to transform possibilities into options, and employed evolutionary algorithms to turn those options into solutions. Our grand unveiling at The Summit for Asset Management in Toronto in 2018 was, from my perspective, an unequivocal hit—thanks to the organizers for giving us a platform to sound smarter than we probably are.

A heartfelt thank you goes to the fictional Department of Offence of Aitopia for a research grant that fueled these early works. The grant was as real

PREFACE

as my ability to juggle flaming torches while riding a unicycle. The research inspired *The Transformable Enterprise: Enterprise Architecture in a New Key*, coauthored with Noah. This book illuminated the clandestine proceedings of the Epoch Making Event, later rebranded as the Epoch Busting Event of the New Key Club—one of the world's most secretive intellectual societies.

Figure 1. *Evolveburgh's transparency initiative rebranded the Department of Defense to the "Department of Offence"—straight to the point. The Library became the "Repository of Already Thought Thoughts," and the Zoo is now the "Enclosure of Reluctant Animal Envoys." Garbage trucks roll out as "Wagons of Past Decisions," and introverts sport "Sporadic Sociability" badges. Evolveburgh, now the City of Silent Introspectors, stands as a testament to the unspoken depth behind every candid word, a sanctuary where wisdom and authenticity are found in the quiet spaces between the lines*

In a twist of fate, the University of Evolveburgh (the second capital of Aitopia) granted me a study leave, even though I never actually worked there. Perhaps they simply wanted to ensure I never showed up! This fortunate turn of events allowed me the time to travel, write, and deeply reflect on the complexities of architecting sociotechnical systems. During this unexpected year of intellectual exploration, I penned *The Triadic Evolution: Reimagine Tomorrow's Sociotechnical World*. This book marks a pivotal shift in our understanding of evolution, explaining how biological evolution, evolution by organization, and evolution by extension converge to empower us to actively shape our destiny. By intertwining these three dimensions—biological evolution, where natural forces shape life; evolution by organization, where structured systems drive progress; and evolution by extension, where technological advancements extend our capabilities—we can actively architect our evolutionary path.

Figure 2. *Evolveburgh stands as an unexpected leader in Aitopia's capital race, propelled by a hundred years of unresolved naming debates. The yet-to-be-named first capital remains unbuilt, but its planners, motivated by the council's lethargic choices, are already drafting for the next Ice Age, featuring high-tech igloos and yeti robots. The name "Precedentia" almost broke the deadlock—until the spelling bee began*

PREFACE

In December 2023, I ambitiously applied for one of 50 OpenAI Superalignment Fast Grants. Despite not being among the lucky 50 out of about 2700 applicants, the experience inspired my latest book, *AI Safety and Security: Architectural Context, Perspectives, and Insights*, published this year. It's a detailed exploration of aligning and securing superhuman AI systems, perfect for those sleepless nights when you're pondering the future of humanity.

This book, *Unit Oriented Enterprise Architecture*, was sponsored by Evolveburgh City Hall under the condition that I include pictures illustrating the fascinating and enigmatic life of Evolveburgh. As one of the most technologically advanced cities in the world and the only one reputed to have had contacts with extraterrestrial beings, Evolveburgh offers a unique backdrop for this work. I extend my deepest gratitude to the Mayor of Evolveburgh for this opportunity.

Figure 3. *Residents of Evolveburgh admire the Mayor's "Zugzwang Urban Design." They're not sure if it's urban genius or just a maze of dead-ends. But in Evolveburgh, a wrong turn might lead to a surprise party or a dimension where socks always match and cats obey commands*

PREFACE

The images within these pages capture the city's blend of futuristic innovation and otherworldly charm, providing a visual journey alongside the conceptual one.

Figure 4. *Evolveburgh: Where intergalactic tourists are kindly advised to explore the next galaxy to maintain the city's unique otherworldly charm*

So, sit back, grab a cup of your favorite beverage, and join me on this whimsical yet informative journey through the realms of unit oriented architecture. As we explore the complex layers of this design philosophy, you'll also discover the charming quirks and hidden gems within the vibrant city of Evolveburgh—a place where tradition and innovation intertwine seamlessly.

Introduction

Before we "Haussmannize" our enterprise architecture, let's remember that sometimes, the best renovations start not with a blueprint, but a bang.

We need to demolish the concept of enterprise architecture as we know it.

It's a bold statement, but history teaches us that sometimes the old must be razed to make way for the new. This was dramatically exemplified by Napoleon III in the mid-19th century when he envisioned a modernized Paris. To bring this vision to life, he enlisted Baron Haussmann, known as the "artist of demolition," who used literal cannons to clear entire neighborhoods, paving the way for broad boulevards and a modern cityscape. Napoleon III's directive was clear: replace the congested and disease-prone medieval Paris with a structured, efficient, and aesthetically appealing urban environment.

Under Napoleon's direction, Haussmann's radical methods and his commitment to transformation turned Paris into a model of modernity replete with wide avenues that improved traffic flow, beautiful buildings that uplifted the city's image, and a vastly improved water and sewage system that addressed critical public health needs. This transformation, though controversial, was fundamentally about tearing down to rebuild better, stronger, and more suitably for the new ages.

Today, we face a paradox. Our task is to dismantle deeply entrenched systems and processes in enterprise architecture—systems not made of bricks and mortar, but of code and convention.

INTRODUCTION

Figure 1. *In a bold stride toward progress in Evolveburgh, where "breaking down concepts" is taken literally and abstract thinking meets concrete action, the citizens are not just destroying notions like "Time" and "Justice"; they're "reimagining" them, one swing at a time. Meanwhile, their mayor, lost in bureaucratic revisions, unwittingly champions the chaos he aimed to control. His bill to prevent conceptual vandalism, twisted by amendments, has become the official playbook for intellectual anarchy—a perfect case study in policy backfire... or an "innovation bonfire."*

Despite its primary aim to outline the components of the enterprise and their interconnections, modern enterprise architecture has faltered. It has devolved into an ongoing, almost Sisyphean task, attempting to harmonize everything from strategy to structure and behaviors—an endeavor as futile as trying to seize the unseizable. As John Zachman observed in his blog, "The Enterprises of today (2015) have never been engineered. They happen... incrementally... over the life of the Enterprise as it grows and requires formalisms... "systems", manual and/or automated." This incremental development, he noted, often results in fragmented structures, making the goal of a cohesive architecture elusive. Similarly, Jay Forrester warned that "Organizations built by committee, by intuition, and by historical happenstance... often work no better than would an airplane built by the same methods."

Yet, these insights appear to have little effect, as modern enterprises continue to evolve piecemeal. Rapid advances in enterprise technologies—cloud, multicloud, data platforms, DevOps, AI, and beyond—move so quickly that few have the time or desire to scrutinize the enterprise as a whole. Instead, the enterprise itself is often perceived as a background environment, like the air we breathe—present, but taken for granted, and rarely examined in its entirety. This lack of holistic scrutiny perpetuates the challenges faced by enterprise architecture.

Furthermore, by breaking down EA into its "constituent" parts—business architecture, application architecture, data architecture, and technology architecture—based on a reductionist approach, the industry has managed to address individual components effectively but fails to reintegrate these into a cohesive whole. This fragmentation leads to a critical oversight: the grand synthesis, the alchemical fusion of these architectures into a harmonious whole, remains unaccomplished, an essential task that remains unrecognized.

While the reductionist method helps us analyze, understand, and evolve individual disciplines, it blinds us to the architectural structure of the enterprise. Thus, as architects and stewards of the whole, we must carefully balance these reductionist insights with an appreciation for emergent complexity. This book proposes starting with the grand vision of

INTRODUCTION

the enterprise and decomposing it into manageable pieces that not only retain the essence of the whole—which is more than the sum of its parts—but also address the critical nuances that make each enterprise unique.

Just as Haussmann wielded demolition to usher in a new era for Paris, we, too, can embrace a transformative fervor to rebuild enterprise architecture for the modern age—without waiting for a "Napoleon" to drive the change. By taking bold steps, we can replace outdated structures and systems with a cohesive, agile framework that better serves today's dynamic environments and future-proofs our organizations against ongoing disruption. Undoubtedly, our challenge of deconstructing the existing enterprise architecture to make room for a novel architectural style—one that ensures adaptability, coherence, and business alignment—is arguably more daunting than Haussmann's task. While Haussmann dealt with the cluttered streets of Paris, we're wrestling with the invisible tangles of digital enterprise architectures. And let's be honest—sometimes, these legacy systems feel as ancient as medieval alleyways. Sure, the challenge is big, but so is the chance to upgrade! By channeling a bit of Haussmann's "creative destruction" (minus the cannons), we can finally bulldoze those outdated systems and build something that doesn't creak when business or technology shifts. Let's pave the way for an architecture that's ready for the digital age—because who doesn't want an enterprise that's more Tesla than horse-drawn carriage?

Join us on this journey to revolutionize your architectural approach, enhance your organization's agility, and achieve unprecedented technological harmony.

Figure 2. *Despite the red tape parade in Evolveburgh, stakeholders continue their magical mystery tour of alchemical fusion, turning business, application, data, and technology architecture into a harmonious "Architectural Smoothie."*

The Cost of Missing Vision: Disjointed Efforts in Enterprise Architecture

Consider a common scenario within enterprise architecture: each architect—whether focused on business, application, data, or technology—often operates with a primary focus on their designated domain. While this compartmentalization allows for deep specialization, it can also lead to fragmentation when individual objectives take precedence

INTRODUCTION

over collective enterprise goals. Without a grand unifying vision, architects may lose sight of how their work aligns with broader architectural picture.

In such environments, business architects may concentrate on aligning strategic objectives with operations, sometimes overlooking technological integration. Application architects, while creating software tailored to specific business needs, may inadvertently create systems that are difficult to integrate across the enterprise. Data architects, tasked with managing robust data systems, might overlook how this data could be leveraged across the organization to drive broader business value. Similarly, technology architects select and implement technologies based on immediate needs without considering long-term adaptability or alignment with other domains.

This example reflects a larger pattern across enterprises lacking a unified vision. It's like attempting to assemble a complex machine without a blueprint—each component may be well crafted, but without a shared understanding of how the pieces fit together, the result is misaligned and lacks coherence. Architects may share a general goal, yet without clarity on how their parts interconnect, conflicting priorities and isolated focus lead to a fragmented outcome.

In reality, the situation is even more complex, reflecting our tendency to abstract away the complexity of the context and take things for granted—much like birds soaring through the sky, unaware of the air that supports them. Each stakeholder, deeply immersed in their own specialized domain, fails to perceive the larger ecosystem that surrounds them. This siloed approach leads to high-quality outputs in isolation, which unfortunately do not add up to a cohesive or fully functional whole. The result is a series of well-intentioned but ultimately disjointed efforts that do not align with the enterprise's overarching goals.

Thus, the absence of a unified architectural vision—not just a strategic vision—breeds inefficiency but also squanders opportunities for innovation and synergy. While traditional EA focuses on aligning strategy, it often lacks a clear, shared understanding of the enterprise's architectural structure—its high-level components and interconnections. A shared vision is essential, not just beneficial, for harnessing the collective strengths of all stakeholders and guiding everyone toward a common goal. This ensures that individual efforts contribute to the success of the whole, avoiding the pitfalls of working in isolation.

Figure 3. *In Evolveburgh, the saying "a camel is a horse designed by a committee" was put on hold when the committee designed a kangaroo instead.*

INTRODUCTION

Blueprint for Success: The Transformative Power of Visionary Enterprise Architecture

Imagine the transformative impact of adopting a visionary architectural style in your enterprise, one that resonates through its simplicity and clarity.

This architecture isn't just a framework; it's a universally understandable metaphor that everyone, from executives to technical staff, can grasp and rally behind. Such clarity not only facilitates more effective communication among stakeholders but also places the architecture itself under a positive pressure for continuous refinement and improvement.

In a landscape where AI safety and security are paramount, this architecture also ensures that every core component is accountable. Clear delineation of responsibilities and functions means that every integral part of the system can be monitored and audited for compliance and performance, thus safeguarding your enterprise against potential vulnerabilities and threats.

Additionally, the architecture champions a purpose driven design of core components. This is not just about building functionality but about crafting purpose-driven digital chains that embody the strategic intents of each unit. Each core component is designed with a specific purpose in mind, which is supported by a coherent chain of constructs that align seamlessly with your business objectives.

This architecture supports easy transformability and reconfigurability. As technological landscapes evolve and societal needs change, your enterprise remains agile and adaptive. This flexibility is essential for maintaining relevance and competitive edge in a rapidly changing world.

Moreover, this approach simplifies the complex dependencies that typically exist between various enterprise components. By streamlining these relationships, the architecture reduces the convoluted and intertwined communication flows that often escalate coordination costs. Simpler communication pathways enhance efficiency and effectiveness, enabling quicker decision making and better alignment of initiatives.

Ultimately, by embracing this innovative architectural style, your enterprise can achieve a level of dynamism and clarity that not only meets current technological and societal demands but also positions you well for future challenges and opportunities. This is the pleasure and power of a well-conceived enterprise architecture—one that every stakeholder understands and contributes to, driving collective success.

Figure 4. *In Evolveburgh, readers take things so literally that when a book says, "Imagine the transformative impact of...," they start wearing hard hats in anticipation of the impending "construction" in their minds.*

INTRODUCTION

Navigating the Blueprint: Key Learnings in Enterprise Architecture

This book is designed to serve as a comprehensive guide to understanding and implementing a revolutionary approach to architecting, universally understood, adaptable, and elastic human-centered large sociotechnical systems. Here's what you will learn through the chapters:

- **Understanding sociotechnical systems:** You will explore the concept of building not just enterprises but large sociotechnical systems that integrate both social and technical elements. The book delves into the principle of Triadic Evolution—biological evolution, evolution by organization, and evolution by extension. You'll understand how these evolutionary processes influence the architecture of sociotechnical systems and how human capabilities in making, computing, and relating play a key role in shaping these systems.

- **Principles of Unit Oriented Enterprise Architecture:** The book introduces and explains the principles of a novel architectural style, Unit Oriented Enterprise Architecture, which is inspired by the fleet metaphor. This metaphor helps to visualize and conceptualize the dynamic and interconnected nature of enterprise components, akin to a fleet of ships moving in coordination toward a common goal.

- **Purpose Driven Design:** You will learn about the principles of Purpose Driven Design, which is fundamental for the core entities—units—of large sociotechnical systems. This approach ensures that each unit's components are aligned with the unit's specific purpose, thereby enhancing functionality and coherence across the enterprise. This alignment facilitates not only operational efficiency but also ensures that each unit contributes optimally to the overarching goals of the organization.

- **Software engineering in the Cloud Era**: A part of this book is dedicated to the often-overlooked aspect of code organization within cloud infrastructures. You will gain insights into organizing code in a way that strictly aligns with the architectural framework, enhancing system clarity, scalability, and manageability. This focus on organizational strategy addresses a critical gap in current practices, shifting the emphasis from merely writing code to strategically positioning it within the architecture.

- **Streamlining enterprise interactions**: Discover a modern approach to managing interactions between enterprise components. The book introduces the principles of indirect coordination, which streamline communication and operations across various parts of the organization. This approach reduces complexity and facilitates more efficient and agile interactions within the enterprise.

- **Preparing for the future of LSS (large sociotechnical systems)**: Finally, explore opportunities for streamlining architecture of large sociotechnical systems. The book examines how enterprises can adapt to evolving sociotechnical landscapes, ensuring resilience and readiness for future challenges and opportunities.

INTRODUCTION

Figure 5. *In Evolveburgh, the enterprise architecture is so avant-garde that local Malevichs and Kandinskys use city data lake blueprints and AI hallucinations as inspiration for bold, chaotic art—proving that in modern art, one person's architecture is another's masterpiece, even if no one knows which way is up.*

By the end of this book, you will have a thorough understanding of how to apply these principles and techniques to enhance the architectural structure and operational efficiency of not only enterprises but any large sociotechnical system. This knowledge will empower you to lead your organization through the complexities of the modern technological landscape toward sustained success and innovation.

Goals and Scope of the Book

The book is designed to serve as a comprehensive guide for understanding and designing large sociotechnical systems (LSS). The primary goals and scope of the book are outlined below to provide readers with a clear road map of what to expect and the insights they will gain:

- **Demystifying LSS**: To clarify the complex nature of LSS, breaking down their components, dynamics, and the interplay between social and technical elements.

- **Providing a framework for design**: To offer a robust framework for the effective design and management of LSS, focusing on integrating sociotechnical elements through Purpose Driven Design principles.

- **Bridging theory and practice**: To bridge the gap between theoretical understanding and practical application, the book offers hypothetical examples and case studies that illustrate how UOEA principles might be applied in real-world scenarios.

- **Emphasizing strategic code organization**: To introduce principles for organizing code within cloud infrastructures to align with the architectural framework, enhancing scalability, manageability, and clarity in the Cloud Era.

- **Streamlining interactions and communication**: To provide a modern approach to managing interactions within LSS, focusing on indirect coordination to reduce complexity, improve agility, and streamline communication across components.

- **Promoting innovative thinking**: To encourage innovative and forward-thinking approaches to LSS, highlighting the latest trends and future directions.

Scope of the Book

- **Understanding LSS**: An exploration of the fundamental concepts underlying sociotechnical systems. It delves into the integration of social, physical, and digital realms, examining the interrelations among LSS components and the various strategies for their interaction. This foundational knowledge sets the stage for a deeper understanding of LSS dynamics.

- **Architectural strategies**: A detailed justification of novel architectural strategies and models used in the design of LSS. These strategies are not just theoretical constructs but are grounded in practical application, offering readers a road map for the effective architectural planning of LSS.

- **Hypothetical case studies and applications**: Hypothetical case studies from various sectors are provided to illustrate the practical application of LSS principles. These case studies shed light on the design and operational challenges inherent in LSS, offering insights into potential solutions and best practices.

- **Methodological approaches**: A novel methodological approach to architecting, implementing, and operating LSS is introduced. This approach is designed to guide professionals and organizations through the complexities of LSS, from conception to operation.

- **Challenges and solutions**: Discussion of common challenges encountered in LSS, such as scalability, integration, and adaptability. It proposes potential solutions and strategies, particularly emphasizing the power of metaphoric modeling as a transformative tool for understanding and resolving these challenges.

- **Future of LSS**: Exploration of emerging trends, technologies, and potential future developments in the field of LSS.

Target Audience

The book is intended for a broad range of readers, including system architects, organizational leaders, IT professionals, policymakers, and students in the fields of system design and organizational studies. It aims to provide valuable insights for anyone involved or interested in the planning, implementation, and management of complex sociotechnical systems.

The book aims to equip its readers with a deep understanding of LSS, practical tools for their design and management, and an appreciation of the evolving landscape of sociotechnical systems. Through this book, we endeavor to contribute to the advancement of knowledge in this vital field, empowering professionals and organizations to build more effective, resilient, and adaptive systems.

Methodological Approach to LSS

The book presents a comprehensive methodological approach to architecting large sociotechnical systems, encompassing a series of interconnected stages from conceptualization to operation.

INTRODUCTION

- **Rethinking enterprise architecture through LSS**: The approach in this book stems from a critical reassessment of traditional enterprise architecture principles. Recognizing the prevalent issues in current enterprise architecture practices—issues that are often overlooked by both the industry and practitioners—we propose a more expansive and general perspective: the architecture of large sociotechnical systems. This shift in focus broadens the scope from conventional enterprise-specific frameworks to a universal design methodology that can be applied to diverse and complex sociotechnical systems. This reconceptualization aims not only to address the underlying problems in enterprise architecture but also to provide a more holistic and adaptable framework suitable for the complexities of LSS.

- **Theoretical foundations**: Central to this approach are novel theoretical constructs that form the backbone of the methodology. These include the introduction of three types of evolution: biological evolution, evolution by organization, and evolution by extension. These concepts provide a framework for understanding the dynamic nature of LSS. Additionally, the methodology introduces universal and uniform building blocks for LSS units: legitimate, purposeful, and accountable social components are extended by technology. This theoretical grounding ensures a robust and adaptable foundation for LSS design.

- **Metaphorical modeling**: The model phase involves the decision to design LSS based on metaphorical modeling mechanisms. This innovative approach involves identifying suitable metaphors that resonate with the characteristics of LSS, aiding in the conceptualization and communication of complex ideas and structures in a more relatable and understandable manner.

- **Architectural strategy—Unit Oriented Enterprise Architecture**: The architectural phase introduces a novel strategy called Unit Oriented Enterprise Architecture. UOEA is an approach specifically tailored for designing LSS, focusing on the modular and interconnected nature of sociotechnical units. This architecture emphasizes the importance of coherence and synergy between the constituent units of an LSS.

- **Implementation strategy—balancing organization and engineering in LSS**: In the implementation phase of designing LSS, our approach distinctly addresses two fundamental aspects: organization and engineering. While engineering pertains to the development and technical integration of applications and suites of applications, the organization aspect is particularly critical, focusing on the strategic arrangement of enterprise software artifacts within cloud environments. This organizational strategy emphasizes aligning units' artifacts with the resource hierarchies offered by cloud providers, such as GCP's four-level hierarchy (Organization, Folder, Project, and Resource).

  This emphasis on organization involves meticulously structuring and managing these artifacts across different cloud regions to ensure sociotechnical alignment and optimize performance, compliance, and accessibility. For example, in GCP, an enterprise might map its divisions and departments within the "Organization" and "Folder" levels, respectively, ensuring that the resources are logically organized in alignment with the enterprise's internal structure. Similar strategies can be applied in other cloud environments like AWS and Azure, each with their unique resource hierarchy.

This dual approach of organization and engineering in the implementation phase ensures a comprehensive and coherent development of LSS. By properly organizing the enterprise software artifacts in the cloud, the implementation not only meets the technical and functional requirements but also aligns with the enterprise's geographical and structural nuances. This balance is vital for the successful deployment and operation of LSS, ensuring they are scalable, adaptable, and aligned with the overarching organizational strategy.

- **Operations strategy—effective coordination mechanisms**: Finally, the operations strategy addresses the interactions of sociotechnical units within the LSS. This involves employing effective coordination mechanisms that leverage both indirect and direct communication patterns. Such mechanisms are important for maintaining the operational integrity of LSS, ensuring efficient and harmonious functioning of the various interconnected components.

Overall, this methodological approach to architecting LSS offers a structured yet flexible framework, guiding practitioners through every stage of LSS development. From the initial conceptualization to implementation and coordinated interactions, this methodology provides a comprehensive toolkit for effectively designing and managing large sociotechnical systems.

Structure of the Book

In the chapters that follow, we will delve deep into the problem, idea, theory, model, design, implementation, and interactions of large sociotechnical systems, using enterprises as our primary examples. By the end, readers will be equipped with the tools, knowledge, and inspiration to envision and craft enterprises that are not just functional but exemplary in their design and operation.

Chapter 1: Problem

Before diving into the complexities of modeling and the transformative power of metaphoric representations, it's imperative to understand the genesis of the challenges surrounding large enterprises. Chapter 1 sheds light on this foundational issue.

Enterprises, as they stand today, are not the result of meticulous planning or architectural prowess. Rather, they've emerged through a series of incremental developments, shaped by myriad forces and decisions over time—"by committee, by intuition, and by historical happenstance" using the words of Jay Forrester who is known for his work in system dynamics and for exploring how complex systems evolve over time. The landscape of these behemoths is similar to a cityscape filled with structures of varying utility—many unfinished, some functional yet inefficient, and others outright abandoned. If we were to visualize our software and technological ecosystems in the same way we perceive physical infrastructures, the scene, as eloquently described by Marc Sewell and Laura Sewell in *The Software Architect's Profession: An Introduction*, would be one of stark contrasts—grand skyscrapers standing next to derelict buildings, with both old and new projects helmed by the same teams that led previous ventures astray.

INTRODUCTION

This chaotic landscape isn't merely a result of historical happenstance but is intensified by the cacophony of voices. Numerous stakeholders, each with their own priorities, vie for influence. Architects, too, are not immune to this discord, often having conflicting visions and priorities. Within this noise, a pressing question emerges: What does the enterprise truly look like? More importantly, who truly needs it?

The current state of affairs reveals a disturbing reality—there's neither a bottom-up push from the industry nor a top-down directive from the enterprise itself. While brilliant minds are engrossed in leveraging cutting-edge technologies, others are swayed by industry trends, often prioritizing buzzwords over tangible outcomes. The enterprise, much like water to a fish, remains invisible to most. This oversight isn't merely due to ignorance but stems from a conscious choice. After all, incorporating enterprise architecture is a complex endeavor that demands time, resources, and effort.

The ramifications of this neglect are palpable. Enterprises remain rigid, unable to adapt seamlessly to external changes, be it mergers, market shifts, regulatory updates, or internal reorganizations. Transformation becomes an uphill battle, fraught with challenges.

It's against this backdrop that the core idea of this book emerges. We aim to pivot from the quagmire of traditional enterprise architecture, which continually finds itself disoriented in a real-world maze, and shift our focus toward the broader discipline of architecting large sociotechnical systems. This monumental shift demands a departure from past naiveties, a critical reassessment of prevailing theories, and the adoption of innovative modeling and engineering strategies.

Chapter 2: Idea

Chapter 2 delves into the foundational concepts behind the unit-oriented approach to architecting complex sociotechnical systems, discussing each concept in reverse order as mentioned in the book's title. This unique structure provides a layered understanding, building from the broad implications of system design down to the specifics of unit orientation.

INTRODUCTION

The chapter begins by addressing the significance of systems thinking in enterprise architecture. It argues for viewing the enterprise as a cohesive whole rather than a mere collection of parts. This perspective is critical for identifying underlying patterns and structures that are not apparent when focusing solely on individual components, thus enabling more strategic planning and informed decision-making.

Following the systems approach, the discussion shifts to the concept of sociotechnical systems. It emphasizes the symbiotic relationship between social systems and technological systems. This section highlights how the effectiveness of any significant system relies on this integration, making it essential to consider both human and technological aspects harmoniously.

The narrative then explores why this approach is particularly effective for large entities. It discusses the scalability of the principles laid out in the book, which are applicable not only to small startups and medium enterprises but are especially adept at addressing the challenges faced by large, complex organizations. This scalability is vital in today's globalized business environment, where companies often span multiple geographical and cultural boundaries.

Next, the chapter advocates for a focused approach to architecture that emphasizes the core elements of enterprise architecture: components and their linkages. This method sets aside less central aspects traditionally entangled with enterprise architecture, such as aligning IT with business goals or optimizing resources, to concentrate on designing robust, adaptable, and clear structural systems.

Finally, the chapter concludes with a detailed explanation of why unit orientation is a distinctive and effective approach in the design and management of sociotechnical systems. Unlike traditional methods that may focus on individual parts or modules, unit orientation views the system as a collection of uniformly constructed units. Each unit, comprising integrated social, physical, and digital elements, is designed to function both independently and synergistically within the larger system.

INTRODUCTION

This chapter sets the stage for a profound transformation in the field of enterprise architecture. By proposing a shift from traditional, strategy- and governance-focused architectural practices to a genuinely architectural approach, the book champions a radical but necessary rethinking of how complex sociotechnical systems are conceptualized and constructed. It argues for an architecture that is not only universally applicable across various scales and contexts but also fundamentally capable of addressing the dynamic challenges of modern enterprises designed and operated as large sociotechnical systems. This shift is not merely theoretical but is intended as a practical guide to designing systems that are resilient, adaptable, and deeply integrated with both human and technological capabilities.

Chapter 3: Theory

Chapter 3 deepens the discussion initiated in the previous chapters by exploring the foundational theoretical concepts essential for understanding the unit-oriented approach to architecting complex sociotechnical systems. Presented in an accessible, nonscientific manner, this chapter makes intricate theories comprehensible to a broad audience, ensuring that readers from various backgrounds can grasp and apply these concepts to real-world challenges.

First, it introduces Triadic Evolution as a pivotal concept, illustrating how the intertwined development of biological, organizational, and technological processes drives the evolution of complex systems. It emphasizes the synergy between human capabilities, social structures, and technological advancements, showcasing how these elements co-evolve to enhance the complexity and functionality of sociotechnical systems. This understanding is important for designing systems that are not only technologically advanced but also harmoniously integrated with human and social factors.

Building on the idea of Triadic Evolution, the next part of the chapter focuses on sociotechnical systems. It discusses the importance of viewing these systems as composed of both integrative and operational sociotechnical entities that collaborate to ensure the system's overall effectiveness and efficiency. The key insight is the critical need to design systems with components that are legitimate, purposeful, accountable, capable, and high-performing, working together to achieve organizational goals seamlessly.

Next, Evolutionary Units are presented as the fundamental building blocks of sociotechnical systems. Characterized by a robust social core and supportive technological peripheries, these units ensure optimal function and adherence to ethical standards. This section highlights the importance of maintaining a strong social core, balanced with effective technological enhancements, to sustain the ethical integrity and functionality of sociotechnical systems.

The final key concept discussed is coordination, identified as the essential mechanism for enabling smooth and efficient interactions within sociotechnical systems. Advances in technology, such as AI and machine learning, are transforming traditional coordination mechanisms, making them more dynamic and capable of handling complex operational demands. This shift is essential for modern systems to remain responsive, adaptable, and effective in a rapidly evolving environment.

Chapter 3 not only elucidates these theoretical concepts but also demonstrates how they interlock to form a comprehensive framework supporting the effective design and management of complex, adaptive systems. Each concept builds upon the others, culminating in a robust approach that aligns architectural design with both human needs and technological advancements. This theoretical foundation is essential for anyone involved in the design and management of modern sociotechnical systems, providing the tools and insights needed to navigate and leverage the complexities of such systems effectively.

INTRODUCTION

Chapter 4: Model

While understanding the enterprise from a strategic vantage point is vital, it is equally important to have a robust representation of it—a model that not only depicts but also aids in comprehending the complexities of large sociotechnical systems. This brings us to Chapter 4, where we dive deep into the realm of representative modeling, understanding its vital role in reimagining these immense systems.

RIG Hughes' framework forms the foundation of our exploration. Hughes emphasized the power of denotation, demonstration, and interpretation as core constituents of representation. By establishing a representative relationship, we can probe the internal constitution of our model, delving into its nuances (demonstration), and then translate these insights into the larger target system. This process is not just theoretical; it's a tangible way of comprehending the sophisticated web of LSS.

But what kind of model best aids our understanding? We explore various modeling approaches—singularized, composite, cohesive, and metaphoric—and argue for the unparalleled potency of metaphoric modeling. Metaphors, as expounded by thinkers like Lakoff and Johnson, don't merely describe; they shape our understanding, bridging reason and imagination in a dance of cognitive coherence. They offer a structure—a familiar framework—whereby we can grasp the elusive.

In the vast expanse of LSS, there are elements beyond our immediate perception—too vast, too intricate, or too abstract. Metaphors emerge as our guiding light in such scenarios, helping us visualize the invisible, fathom the unfathomable, and simplify the complicated. They act as cognitive bridges, drawing from familiar, concrete domains to shed light on the more ambiguous and intricate ones. This isn't just a rhetorical exercise; it's a fundamental cognitive process that allows us to bring the vastness of LSS to a "human scale," making them comprehensible and manageable.

Yet, for a metaphor to be effective, it must adhere to certain criteria. It should be relevant, universally understood, specific enough for our purpose, rich in its layers of meaning, and of high quality, especially when representing something as sophisticated as an LSS. In our quest for the perfect metaphor, we aim for one that not only meets these criteria but also resonates deeply with the digital facets of LSS.

As we progress through this book, readers will come to appreciate the power of metaphoric modeling as a transformative tool, reshaping not just how we perceive but also how we design and implement large sociotechnical systems.

Chapter 5: Architecture

Chapter 5 delves into a comprehensive architectural approach tailored for designing and managing expansive sociotechnical systems. Central to this strategy are two pivotal concepts: the comprehensibility of the architecture itself and the changeability of the LSS as a whole. This chapter methodically breaks down the architectural process into its fundamental components, offering readers a detailed road map for constructing and maintaining effective and adaptable LSS.

We begin by discussing the component selection patterns of LSS architecture, exploring options such as homogeneous, heterogeneous, and hybrid configurations. This sets the foundation for understanding how different components can be optimally selected and combined to form a coherent and efficient system.

Structural patterns form the next focal point, with an analysis of architectures, including bus, ring, star, tree, mesh, and various hybrid forms like the extended star and snowflake. These structures are important for determining how different components of an LSS communicate and interact, influencing the overall efficiency and resilience of the system.

INTRODUCTION

The chapter then progresses to relational patterns, covering exchange, organizational, collaboration, cooperation, and compliance relations. This section underscores the importance of inter-entity relationships and how they can be effectively managed within a large-scale system.

Compositional patterns such as distributed, partitioned, interlocked, and fused are also examined, providing insight into how various elements within an LSS can be organized and integrated.

A key part of the chapter involves selecting a large enterprise with a geographical divisional structure as a prototype for LSS architecting. This practical example serves as a foundation for relating LSS to maritime formations, where the metaphor of an armada is used to illustrate the dynamics of global enterprises. This analogy extends to discuss regional fleets within the armada, including the roles of flagships, command ships, and operational ships.

The introduction of Unit Oriented Enterprise Architecture is a highlight, presenting a novel architectural style specifically designed for LSS. The chapter outlines the key definitions, core principles (general, structural, and interaction), and the nuances of implementing UOEA in a real-world context.

The benefits of UOEA are then elaborated upon, followed by hypothetical yet applicable case studies and applications that demonstrate the real-world effectiveness of these concepts. These examples provide tangible insights into how UOEA can be applied across various industries and scenarios.

The chapter concludes with final thoughts about integrating UOEA into an LSS strategy, reflecting on how this approach can revolutionize the design and management of large sociotechnical systems. This chapter, building on the foundational concepts established in the previous ones, equips readers with the knowledge and tools to not only comprehend but also adeptly architect vast and complex sociotechnical systems.

Chapter 6: Design

In Chapter 6, we explore the complexity of crafting individual units within a large sociotechnical system. This chapter delves into the unit design vision of UOEA, which employs a bifocal design approach fundamental to achieving a balanced and effective LSS. On one side of the spectrum, we emphasize structural uniformity, where all units share a common structural blueprint composed of digital tools, including apps, applications, and suites of applications. This uniformity facilitates coherence and integration across the LSS.

Conversely, the principles of Purpose Driven Design (PDD) tailor each unit to meet specific goals and stakeholder needs. While the uniform structure provides a shared skeleton, each unit's distinct functions form the flesh and muscle, ensuring that every unit is uniquely equipped to fulfill its role within the LSS. We examine how PDD integrates various constructs—purpose, function, process, structure, order, culture, memory, and storage—into a coherent digital chain that equips each unit to pursue its purpose effectively.

The chapter further details the networked and systemic nature of LSS units, where each unit consists of suites of applications, each suite comprises applications, and each application consists of apps (microservices). In this hierarchy, an integrative component coordinates the others, providing public APIs and user interfaces to ensure seamless operation.

We also discuss the seven core principles of Purpose Driven Design, focusing on purposefulness, modularity, clarity, and integrative functionality. These principles guide the creation of systems that are not only technologically robust but also socially pertinent and organizationally efficient. By balancing uniformity with purpose driven customization, we ensure that each unit within an LSS operates cohesively while being tailored to its specific operational context.

INTRODUCTION

Chapter 7: Implementation

In Chapter 7, we consider the vital aspects of bringing these systems from conceptual frameworks to operational realities, concentrating primarily on Organization and Engineering. This chapter navigates through the nuances of implementing LSS, particularly emphasizing how they are structured and managed within the realms of cloud and multicloud environments.

The narrative begins by exploring the organization of LSS, where we discuss the importance of distributed and partitioned compositions at the system and unit levels. We consider an LSS as a distributed system, essentially a collection of independent units located across various cloud regions or even spanning multiple clouds. At the unit level, the focus shifts to the principle of partitioned composition. Here, the units are organized within their specific cloud environments, ensuring that each maintains its distinct operational boundaries and capabilities while still being part of the larger enterprise ecosystem.

Transitioning to the aspect of Engineering, the chapter provides insights into the development and integration of the unit's building blocks—the applications. It is here that the engineering process in Unit Oriented Enterprise Architecture is clarified, highlighting the creation of applications that align with the unit's defined purpose and function, ensuring their effective contribution to the overarching goals of the LSS.

The chapter also addresses the practical aspects of organizing LSS and units in cloud environments. It covers a range of considerations including geographical, departmental, environmental, and project-based factors, along with a structured approach to naming and tagging. This section is vital for building an agile, compliant, and strategically aligned cloud infrastructure. Further, it delves into the specifics of organizing in various cloud environments like AWS, Azure, and GCP, each offering unique structures and features for optimal organization of LSS.

INTRODUCTION

In discussing the uniform structure of units, the chapter emphasizes that units within an LSS consist of applications or suites of applications. This uniformity is pivotal for ensuring seamless integration and operation within the larger system. Additionally, the chapter touches upon the general considerations for multicloud approaches, highlighting both the benefits and challenges of managing LSS across diverse cloud platforms.

Overall, Chapter 7 stands as an essential guide for those involved in designing and managing LSS. Bridging the theoretical with the practical, it equips readers with the necessary strategies and insights to effectively implement and manage large sociotechnical systems in the modern, cloud-centric technological landscape. The chapter is a narrative journey through the complexities and practicalities of LSS implementation, ensuring a comprehensive understanding for a wide range of readers, from system architects to organizational leaders.

Chapter 8: Interactions

Chapter 8 takes a deep dive into the operational heart of large sociotechnical systems, unraveling the complexities of how these systems function through a division of work and responsibilities among their constituent sociotechnical units. This exploration is not just about task allocation; it's about understanding the strategic orchestration that underpins the effective functioning of the entire system.

At the center of LSS operations are the varied interactions that define the relationships between different units. The chapter explores how these interactions, whether based on exchange, organizational, collaboration, cooperation, or compliance, necessitate tailored coordination mechanisms. It delves into the realms of both direct and indirect coordination, each serving distinct but complementary roles in the operational integrity of LSS. Direct coordination is portrayed as the backbone of structured command-and-control mechanisms, essential for clear and efficient process management. In contrast, the concept of

INTRODUCTION

indirect (stigmergic) coordination introduces a more fluid approach, similar to identifying natural paths or "worknets" in the system's workflow. This approach captures the essence of flexibility and adaptability in operational processes, much like finding and utilizing desire paths in a landscape.

The chapter also highlights the pivotal role of communication patterns in enabling these coordination mechanisms. It illustrates how the effective use of communication, be it direct or indirect, maintains coherent and synchronized operations among the interconnected components of LSS.

In essence, Chapter 8 of the book is a narrative that weaves together the various strands of operational strategy within LSS. It provides a comprehensive understanding of the dynamic mechanisms that drive these complex systems, spotlighting the sophisticated interplay of coordination, communication, and connectivity in ensuring their operational success. This chapter is not just an overview; it's a guide to the pulsating operational life of large sociotechnical systems.

Figure 6. In Evolveburgh, enterprise architects embraced the "Anything as a Silver Bullet" mindset so fully that the silver bullet became an official part of their modeling notation. At city Architectural Review Board meetings, if an architecture is presented without a silver bullet, you'll likely be asked if you've mistaken it for a werewolf summit.

Final Remarks

Given the current state of the world, it is imperative to reemphasize the profound significance of large sociotechnical systems in the growth of modern civilization. Large sociotechnical systems are indeed the

INTRODUCTION

backbone of our society, underpinning the functionality of key sectors and influencing diverse aspects of social, economic, and environmental life. The effective understanding, design, and management of these systems are not just academic pursuits; they are imperative for the sustained progress of society, the fostering of economic development, and the enhancement of the quality of human life on a global scale.

Rooted in the rich history and principles of sociotechnical theory, this book traverses the journey of LSS from their conceptual foundations to actionable frameworks. Building on these foundational concepts, it introduces Unit Oriented Enterprise Architecture, a reoriented approach designed specifically for LSS. UOEA provides a practical and cohesive framework for integrating social and technical components, emphasizing the primacy of social constructs and enabling flexible, scalable, and universally comprehensible designs. This approach is not just a theoretical repositioning; it is a pragmatic strategy for enhancing the efficacy, resilience, and adaptability of these systems in the real world.

The overarching goal of this book is to provide readers deep understanding of LSS, explaining their complexities and presenting UOEA as a cutting-edge methodology for their successful implementation. Whether you are a system architect grappling with the technical workings of LSS, an organizational leader strategizing their effective deployment, a policymaker shaping the regulatory landscape, or a student eager to explore the sociotechnical systems theory, this book is designed to be your guide and companion. It aims to equip you with the critical knowledge and tools required to navigate, contribute to, and shape the evolving domain of large sociotechnical systems.

In summary, this book is more than just an academic treatise; it is a call to action for all those involved in shaping the future of our sociotechnical world. It is an invitation to engage with, contribute to, and shape the narrative of LSS—systems that are pivotal to our collective future and the continued advancement of global society.

CHAPTER 1

Problem

Are enterprise architects secretly lost and just pretending to know what's going on? The structure of the industry seems as fragile as a house of cards in a storm, with megavendors luring architects into their grand illusions. The essence of an enterprise is as elusive as a mirage. Architects may not recognize it even when it's right before their eyes, yet they might find humor in the adage that a camel is but a horse designed by a committee. Some whisper that if even the almighty Kubernetes can't fix it, then perhaps it's a puzzle meant to remain unsolved.

In a world transformed rapidly by digital advancements, enterprises serve as significant markers of human creativity and progress. These entities are dynamic, continuously adapting to the shifts in the digital environment. However, despite their extensive impact and strong presence, enterprises often operate within a complex network of details that are not fully understood, functioning in areas that occasionally exceed the understanding of the individuals who shape them.

In this context, many enterprises lack a cohesive grand vision, often shaped by intuition and historical happenstance (Forrester) rather than deliberate strategic design. This shortcoming arises from the industry's struggle to provide a unified vision amidst the complexity of enterprise technologies and the constant emergence of new tools and trends. Consequently, organizations frequently react to technological shifts instead of being guided by a clear forward-thinking strategy. This often results in a patchwork of systems and practices that, while functional, do

CHAPTER 1 PROBLEM

not necessarily align with a unified purpose or adapt optimally to new challenges. The absence of a clear, overarching vision leads to missed opportunities and inefficiencies, as organizations struggle to pivot or scale in coherent and purposeful ways. Addressing this gap requires a conscious effort to craft a vision that not only reflects the realities of the digital age but also anticipates future directions and innovations, ensuring that enterprises are not merely reacting to changes but are actively shaping their destiny.

Figure 1-1. At the annual Evolveburgh Gala, the architectural designs are not just grand—they're so socially significant they RSVP with a plus one, bringing their lesser-known, but equally ambitious, blueprints to the party

CHAPTER 1   PROBLEM

Why Everything Is Right and Everything Is Wrong (Simultaneously)

In enterprise architecture, there's a paradox. On the surface, everything seems correct, but underneath, there are many problems. It's a notion where everything **seems** to be simultaneously correct and incorrect, a tension that defines the landscape architects navigate. On the surface, the structures of modern enterprises appear sturdy and well crafted, a testament to the diligent efforts of architects striving to ensure endurance and evolution. Yet, beneath this layer of correctness lies a less optimistic reality—a landscape fraught with complications, such as legacy systems that are costly to replace, fragmented data silos impeding seamless integration, and ever-evolving security threats. Architects must contend with the constant challenge of maintaining stability while adapting to rapid technological change, leading to makeshift solutions that temporarily address issues without resolving their root causes. This paradox, where everything appears right and wrong at once, encapsulates the essence of enterprise architecture, challenging architects to navigate ambiguity and uncertainty with finesse.

Initially, architects diligently constructed the sophisticated frameworks of contemporary enterprises, employing a reductionist approach that delineated four core domains: business, application, data, and technology architecture. This method, rooted in the philosophical principle of deconstructing complex systems into more manageable components, served as the foundational framework during the 1990s. However, as digital disruption accelerated, introducing cloud computing, microservices, and agile practices, the reductionist model was augmented to manage new layers of complexity and interdependency across domains. While agile practices improved efficiency through time-boxed iterations, the emphasis on these often architecturally stripped short development cycles frequently compromised the coherence of enterprise systems.

CHAPTER 1   PROBLEM

Cloud workloads and microservices further decomposed enterprises, leading to "Death Star" architectures—highly intricate and interconnected systems (see, e.g., "Webinar: How to Avoid Building a Microservices Death Star" by Software AG, as well as the widely available diagrams depicting the architectures of Netflix, Amazon, and Twitter found across the Internet). This evolution reflects a shift toward integrating rapid change and innovation into the original framework, enhancing but not abandoning reductionist principles.

Building on the foundations of reductionism, contemporary enterprise architects navigate a terrain that has been further decomposed into specialized subdomains beyond the core areas of business, application, data, and technology architecture. This evolution reflects a deeper breakdown into areas such as solution architecture, infrastructure architecture, integration architecture, and digital architecture. While these domains are technically subsets of the traditional areas, they underscore the increasing complexity and interconnectedness of modern enterprise systems, emphasizing bespoke solutions, robust IT management, seamless system integration, and digital innovation. Concurrently, the emergence of information security architecture highlights the critical need for data protection within these frameworks.

The nature of this problem is well articulated in *Functional Reactive Programming* by Stephen Blackheath and Anthony Jones:

> "Software development is a form of engineering, and engineering is based on the philosophy of reductionism. This methodology is powerful; it has been enormously successful at providing us with technology that has transformed the way we do almost everything.

CHAPTER 1   PROBLEM

The reductionist approach to engineering has four steps:

1. Start with a complex problem.
2. Break the problem into simpler parts.
3. Solve the parts.
4. Compose the parts of the solution into a whole.

> Step 4 is where we get into trouble. Reductionism has a hidden assumption of compositionality: that the nature of the parts doesn't change when we compose them. When this isn't true, we can fail badly. If we wrongly assume compositionality, then we have committed a logical fallacy called the fallacy of composition."

CHAPTER 1   PROBLEM

Figure 1-2. *Welcome to Evolveburgh, where our enterprise architecture is like a gripping detective story—always one step ahead! Just when you think you've caught up, it pulls a plot twist that would make Sherlock Holmes scratch his head. It's all fun and games until our crime boards start resembling UML diagrams, and our suspect profiles look like they need their own decryption key*

Yet, beneath this semblance of progress lies a less sanguine reality. The digital structures engineered to underpin business operations are showing signs of strain, often patched together with temporary solutions. Despite the evolution of enterprise architecture, systems remain fragmented, disconnected from broader organizational objectives, and ill-equipped to pivot in response to changing circumstances. While reductionism offers analytical rigor, it falls short of capturing the holistic essence of organizational architecture, resulting in inefficiencies and suboptimal outcomes.

CHAPTER 1   PROBLEM

Enterprise architects, frequently at the forefront of ambiguity, contend with a rigid approach that often prioritizes individual systems over architectural cohesion, leading to a failure to compose these parts into a unified whole. This failure creates a gap when architecture is misaligned with organizational strategy. While effective enterprise architecture should ideally bridge this divide, in practice, fragmented approaches and operational silos hinder the composition of integrated solutions, highlighting the need for more holistic strategies. Through introspection and innovation, architects can transcend reductionism, fostering agility, alignment, and resilience amidst the tumult of digital disruption. The journey of enterprise architecture balances a delicate equilibrium between endurance and evolution. While past methodologies have laid a solid groundwork, the path forward necessitates a paradigm shift—a departure from reductionism toward holistic and adaptive strategies. As architects grapple with the complexities of the digital era—marked by emerging technologies like artificial intelligence and machine learning, evolving business needs driven by market dynamics, challenges in integrating diverse legacy and modern systems, increasing data privacy and cybersecurity demands, and the push for sustainable and ethical practices—their capacity to navigate uncertainty and embrace change will undoubtedly shape the trajectory of enterprises in the foreseeable future.

The Grand Ballet of Stakeholders: A Problem That Can Paralyze

Enterprise stakeholders—each with their unique priorities and goals—often find themselves at odds, leading to conflicts as they vie for prioritization of their interests. In the realm of digital enterprises, these conflicts assume heightened complexity. The size, intrinsic complexity, and invisibility of digital enterprises pose challenges for stakeholders to fully comprehend the implications of their demands. The fear of ending up with a

CHAPTER 1 PROBLEM

"camel" when aiming for a "horse" is prevalent. However, within the context of complex digital enterprises, even achieving a semblance of coherence similar to a "camel" can be deemed acceptable in the absence of a clear "horse."

The dynamic between business and IT sponsors of EA mirrors that of an orchestra with multiple conductors, each representing the interests of their respective units.

Figure 1-3. *At the Evolveburgh Opera House, the symphony of enterprise architecture descends into delightful chaos: multiple conductors battle for the baton, developers perform impromptu solos on instruments they've never played before, and project managers give standing ovations at random—meanwhile, the servers haven't just left the building; they've started a rival jazz band across the street*

In this analogy, nobody assumes responsibility for architecture of the whole enterprise, similar to the orchestra, where interpretations of the music sheet, or enterprise architecture, vary. Attempting to guide the organization, or orchestra, according to their interpretation often results in cacophony rather than symphony, emphasizing the need for unified direction.

Figure 1-4. *In the grand theater of Evolveburgh, the "Stakeholders" ballet unfolds with the coordination of a toddler-led furniture assembly team. Dancers whirl in every direction, each lost in their own rhythm, much like when you're trying to multitask and end up doing nothing. Meanwhile, managers navigate through a digital storm as if steering ancient ships, desperately searching for a GPS route to a mythical digital paradise where the cloud storage is always full*

CHAPTER 1 PROBLEM

Consider the scenario of a multinational corporation integrating digital systems post-acquisition. The involvement of various business and IT units, each with distinct priorities and goals, highlights the challenge; absent centralized architecture management, fragmentation, and lack of coordination lead to redundancies, inefficiencies, and disjointed digital architecture. Competing interests among units breed conflicts, delays, and cost overruns, undermining effective integration and strategic alignment. The absence of a clear strategy and decisive leadership culminates in a patchwork of systems failing to support business objectives adequately, jeopardizing operational efficiency, and impeding synergies from the acquisition.

CHAPTER 1   PROBLEM

Figure 1-5. Disaster struck at Evolveburgh's grand premiere of the "Stakeholders" ice show, the Synchronized Edition of the "Stakeholders" ballet. On stage, the performance was so unexpectedly perfect; it was like watching a group of ducks perform Swan Lake. The skaters moved with the puzzling grace of cats solving Rubik's cubes, leaving the audience quacking in surprise. Mostly comprised of relatives lured in with the promise of free popcorn and distant cousins granted discounts visible from Mars, the spectators responded with a cacophony of boos and whistles demonstrating that, sometimes, flawless execution is simply too much for some to handle

Competing stakeholders can paralyze decision-making in EA, stalling projects and deferring essential decisions indefinitely, resulting in missed opportunities and stagnation. The impact on EA can be profound, as

CHAPTER 1   PROBLEM

it serves as the blueprint for the organization's digital structures and business processes. When stakeholders fail to agree on this blueprint, fragmentation, inefficiency, and ineffectiveness ensue, compromising organizational agility and competitiveness.

Stakeholder conflicts also impede the design and implementation of EA, yielding designs attempting to appease all parties but satisfying none. During implementation, resource allocation discrepancies among stakeholders lead to delays and cost overruns, further exacerbating inefficiencies and operational disruptions.

The potential consequences for an organization of a poorly designed and implemented EA are disastrous. It can lead to inefficiencies, escalated costs, and missed opportunities, thereby jeopardizing competitiveness and long-term viability. Managing the competing priorities of stakeholders in EA demands strong leadership, transparent communication, and a readiness to make tough decisions preventing the paralysis that can result from attempts to appease all parties.

However, it's equally important to ensure that influential stakeholders, who are responsible for the architecture of the entire enterprise, are consistently involved in all high-stake projects and initiatives. Their active participation and the opportunity to voice their opinions can significantly influence the direction and success of these important endeavors. This consistent involvement ensures that the enterprise architecture aligns with the organization's strategic objectives and mitigates the risk of costly misalignments or oversights.

By effectively resolving competing priorities and fostering active involvement from key stakeholders, organizations can navigate the complexities of EA and steer toward a future of growth and innovation.

CHAPTER 1   PROBLEM

The Great Enterprise Mystery: Who Knows What It Looks Like?

In the digital world, there's a paradox: even though humans build modern businesses, these enterprises are often too complex for us to fully understand. Interestingly, these enterprises don't materialize all at once. They are built incrementally, mainly on a project-by-project basis. As a result, the overall enterprise that emerges isn't always a product of intentional design, but rather a cumulative outcome of many separate projects and initiatives. This incremental evolution of fragments of the enterprise contributes to the complexity and challenges in fully grasping the structure and behavior of the whole.

This disconnect is not just a theoretical concern but a practical hurdle in our race toward effectiveness and innovation.

This practical hurdle manifests in several ways, particularly when it comes to enterprise transformation. The inability to fully "see" the enterprise, its building blocks and linkages, can significantly impact the effectiveness of reengineering, reorganization, global transformations, and mergers and acquisitions.

In the context of reengineering, the lack of a comprehensive understanding of the enterprise can lead to misguided efforts. Without a clear view of the enterprise's structure and behavior, reengineering initiatives may overlook critical interdependencies, leading to unintended consequences that can disrupt operations and negatively impact performance.

In the context of reorganization, the complexity of the challenge escalates. Reorganization typically necessitates modifications to roles, responsibilities, and reporting structures. However, without a comprehensive understanding of the enterprise's foundational elements and their interconnections, making informed decisions about these changes becomes a daunting task. This complexity is further compounded when considering the alignment of digital structures with organizational changes. The realignment of digital systems and processes to reflect new organizational structures can be a complex and time-consuming endeavor. Missteps in this process can lead to operational inefficiencies, create confusion, and negatively impact employee morale. Therefore, a clear understanding of both the organizational and digital architectures of the enterprise is vital for a successful reorganization.

During global transformations, the intricacies are magnified due to geographical, cultural, and regulatory variances. The challenge lies not only in aligning global strategies with local operations but also in harmonizing organizational and digital architectures across diverse regions. Without a comprehensive understanding of the enterprise, achieving this global alignment can be a daunting task. The inability to fully comprehend the enterprise can result in misalignments, thereby undermining the effectiveness of the transformation. This highlights the importance of a holistic understanding of the enterprise, encompassing both its organizational and digital architectures, for successful global transformations.

In mergers and acquisitions, a comprehensive understanding of the enterprise is critical for due diligence, integration planning, and post-merger integration. Without this understanding, there's a risk of overvaluing or undervaluing the enterprise, overlooking potential synergies, or failing to identify and mitigate integration risks.

Understanding a whole enterprise with all its complexities is a big challenge when we try to transform it. This shows why we need advanced tools, methods, and skills to help us better understand and navigate the complex nature of today's enterprises. It's not just about improving our technical skills but also about building a culture of learning, curiosity, and adaptability to keep up with the fast changes in the enterprise world.

CHAPTER 1 PROBLEM

Figure 1-6. *In Evolveburgh, enterprise architects gather around a mysterious object rumored to be an "enterprise." As confusion reigns and frameworks are consulted, the age-old question resounds: "But what does an enterprise actually look like?" Armed with upside-down blueprints, an intern shares a legend: the last person to truly understand the enterprise was a retired employee who now spends his days in a quiet corner of the library. He once said, "The enterprise is not an object to be seen, but a vision to be shared." Enlightened, the architects decided that the enterprise is like a doughnut—it's round, it's sweet, and there's a hole in the middle where the budget used to be*

Consider the analogy of constructing a large city in total darkness. We may be familiar with the roads and intersections, but without a bird's-eye view, our understanding of the city's layout remains incomplete. In such a

CHAPTER 1   PROBLEM

scenario, it's challenging to construct the individual buildings. So, we build individual blocks and systems, similar to applications and microservices. However, even when these individual blocks and systems start working and functioning, we often hesitate to assemble them into larger, more complex structures, similar to city buildings.

Figure 1-7. Due to Evolveburgh's peculiar by-laws that permit city construction only in complete darkness, the local Architecture Association had to invent special "night vision tools." This was to avoid the all-too-common scenario where builders, after constructing individual blocks and systems, would scream in surprise upon seeing what they'd built in the light of day. Now, they can scream in real time, right as they're building it

CHAPTER 1   PROBLEM

Just as city planners need a comprehensive plan, so too do architects of digital enterprises. In the digital realm, architects and engineers may understand individual systems, processes, and data flows, but the overall picture often eludes us. This lack of holistic understanding can lead to inefficiencies, redundancies, and missed opportunities for innovation. It can also pose significant risks, as unidentified vulnerabilities may lurk in the unseen corners of the enterprise.

To navigate the future with confidence and truly harness our potential, we, as architects and engineers, must develop effective "night vision tools." These are methodologies that allow us to visualize and comprehend the complex structures of digital enterprises. With these tools, we can illuminate the path toward a more efficient and innovative future.

Complex yet Unperceived Architecture: The Lack of Clear Vision

Every company is different, with its own goals, problems, and operating environment. A successful architectural approach needs to be flexible and able to meet each company's specific needs. Many people use frameworks from experts like Zachman and Spewak. These frameworks focus on understanding a company's specific situation and creating solutions that fit. However, these frameworks shouldn't stop us from making a clear plan for the company and basic designs for building companies.

This might seem to advocate for a more standardized, one-size-fits-all solution, which might not take into account the unique needs and characteristics of each enterprise. However, Albert Einstein's famous saying, "If you can't explain it to a six-year-old, you don't understand it yourself," applies not only to simple concepts like the growth of a plant, more complex concepts like a city, but also to the digital enterprise. This underscores the importance of simplicity and clarity in understanding and explaining complex systems.

17

CHAPTER 1   PROBLEM

In this light, there is a need for a foundational template or blueprint that can serve as a starting point for all enterprises. While this blueprint would be universally applicable, it would also be flexible and adaptable, allowing it to accommodate and address the unique complexities and diversities of each individual enterprise. This does not mean a rigid, inflexible framework, but rather a flexible model that can be adapted to the unique needs and characteristics of each enterprise.

Figure 1-8. In Evolveburgh, even assembling IKEA furniture without a Grand Vision, a template, and a blueprint is considered a heroic feat

Contemporary technology is continually evolving, with various methodologies and tools, including AI-based ones, being developed on an hourly basis. Now, more than ever, there's a need to propose a single "blueprint" for what an enterprise looks like. This blueprint would serve as a guide, a starting point from which each enterprise can develop its own unique architecture.

By having a clear vision and a basic blueprint, we can be constantly working toward developing better ways to understand and design enterprises. This will enable us to navigate the complexities of the digital world, harness the potential of emerging technologies, and drive innovation and efficiency in our enterprises.

The Abstract Nature of Enterprises

Imagining and reimagining digital enterprises is a complex task, fraught with challenges. Despite being surrounded by the manifestations of enterprise architecture, the true nature of the enterprise remains somewhat elusive. This is due to its immediacy, invisibility, size, and the sheer number of people involved in its construction.

The enterprise stands as an abstract entity, a puzzle that seems to lack definitive structure or form. It's like trying to understand the shape of a cloud while being within it. Just as a cloud is vast and continually changing, with its full shape obscured to those within it, a contemporary enterprise is a dynamic, highly fragmented entity that constantly evolves and adapts to the shifting business landscape.

For enterprise architects and other contributors, the enterprise appears as an ever-changing canvas. Automated systems and isolated islands of automation emerge, each forming a part of the larger entity. However, these components are frequently developed in isolation. Each team or individual prioritizes their immediate tasks without a comprehensive understanding or appreciation of the entire enterprise. This isolation often results in a lack of cohesion and integration, with different parts of the enterprise functioning in silos, disconnected from one another.

CHAPTER 1   PROBLEM

Figure 1-9. In the ever-transforming city of Evolveburgh, enterprise architects are as ubiquitous as pigeons in the park, and just as likely to leave their "creations" on your windshield. As a result, the abstract nature of enterprise architecture has led to its surprising reclassification as abstract art. At Evolveburgh's most prestigious auction house, a UML sequence diagram by a local EA recently fetched a stunning $19 million, catapulting the architect into Van Gogh territory. Now, every enterprise architect in Evolveburgh is doodling their way to fame, turning UML diagrams into surrealism, and hoping their next system model might just be their ticket to the Louvre. Welcome to Evolveburgh, where every blueprint could be the next Matisse, every architect an aspiring Andy Warhol, and every meeting room a potential gallery—just mind the "pigeon art" on your way in!

This abstract nature poses significant challenges. Without a clear, overarching vision or understanding of the entire enterprise, strategic decision-making becomes cumbersome. Identifying redundancies or spotting opportunities for innovation is similar to navigating a maze without a map—challenging and often inefficient.

Consequently, there is a pressing need for tools and methodologies that provide a holistic view of the enterprise. Similar to how a map aids in navigation, a clear and comprehensive understanding of the enterprise can guide decision-making, optimize processes, and foster innovation.

While the abstract nature of enterprises presents challenges, it also offers substantial opportunities. By developing a clear vision and comprehensive understanding of the enterprise, stakeholders can navigate the complexities of the digital world more effectively. Harnessing the potential of emerging technologies becomes feasible, driving efficiency and innovation within the enterprise.

This is the exciting challenge that lies ahead for enterprise architects and all contributors in the digital age. Navigating this complexity, transforming abstract concepts into concrete strategies, and guiding the enterprise through the evolving digital landscape are crucial tasks that will define the success of future organizational endeavors.

The Evolving Life Cycle of Digital Environments

Digital environments are not static; they possess a life of their own. They grow, renew themselves, and have the potential to endure for decades. Yet, they lack the time-tested essence seen in physical structures, which often carry a legacy, values, and identities that span generations. Digital enterprises are susceptible to constant change, sometimes losing their original essence in the whirlpool of rapid technological advancements and shifting stakeholder interests.

The ephemeral nature of virtual spaces stands in stark contrast to the enduring solidity of physical edifices. While cathedrals and monuments

CHAPTER 1   PROBLEM

bear the weight of history, digital platforms and solutions often flutter in the winds of innovation, their foundational principles and identity at risk of being eroded by the relentless tide of progress.

Figure 1-10. *In Evolveburgh, where foundations are built on solid stone or fleeting pixels, grand cathedrals and ancient monuments bear the weight of history, while digital platforms disintegrate like cookies in grandma's vintage browser. To combat this ephemeral nature, Evolveburgh's enterprise architects have erected monuments of their own: an 8-meter-high pixel majestically sits in the city's most scenic lake as if it tumbled from the sky, a colossal USB stick stands beside the city hall reminding politicians to fulfill their promises, a towering "Ctrl+Z" symbol in the town square assures citizens that it's never too late to undo, and a grand "404 Error" sign graces the central park—because in Evolveburgh, even the mistakes are monumental!*

This constant state of flux challenges digital enterprises to not only adapt and evolve but to do so without forfeiting their core essence and values. The challenge is similar to rebuilding a ship while at sea, where each plank replaced must be carefully chosen to ensure it doesn't alter the vessel's original character. As such, digital leaders are tasked with the delicate balancing act of embracing cutting-edge technologies and trends while preserving the unique identity and values that distinguish their enterprise, ensuring that amidst the ceaseless waves of change, the soul of the digital entity remains intact, vibrant, and true to its founding principles.

The Need for a Cohesive Strategy

The digital world is in a state of continuous transformation, with large-scale changes driven by emerging technologies and diverse user needs. It is a realm where transformation is both ubiquitous and rapid. Yet, amidst this frenzied pace, there lies an opportunity for patient cultivation of enterprise at the grassroots level, fostering both top-down and bottom-up dynamic transformations. This approach, however, demands a cohesive strategy, a vision that encompasses the various facets of enterprise development, fostering a harmony that currently seems elusive.

To navigate the complexities of today's digital enterprise landscape, a shift in perspective is necessary. We must foster an environment where innovation is guided by a clear, encompassing vision, allowing for growth and adaptation without losing sight of the fundamental values and goals. By acknowledging and addressing the "Great Enterprise Mystery," we pave the way for structures that are not just expansive and flexible but also coherent and purpose driven.

CHAPTER 1   PROBLEM

A Love Triangle: Physical, Digital, and Social, All Wanting Attention

A complex interplay is unfolding among the physical, digital, and social dimensions of contemporary enterprises, each competing for integration and priority. Central to this complex dynamic is the principle of enterprise composability, which involves the configuration of building blocks that are not only legitimate, purposeful, and accountable but also instrumental in forming an entity that reflects these characteristics in its structure and operations.

Figure 1-11. In Evolveburgh, the battle for dominance among the physical, digital, and social dimensions isn't just metaphorical—it's a daily exercise regimen. Today's meeting agenda features a literal tug-of-war: neurons flexing their synapses in strategic maneuvers, bits engaging in intense data crunches, and atoms handling the heavy lifting. And the winner of this multidimensional contest? Well, the results are still loading... please stand by!

The path to achieving optimal synchronization, where all three dimensions merge harmoniously, is still being forged by the software industry. The ultimate objective extends beyond merely creating an efficiently operating enterprise. The aim is to construct a dynamic entity where each component exemplifies a commitment to legitimacy, purposefulness, and accountability, thereby cultivating a holistic environment that resonates with integrity and harmony at every level.

The Interplay of Architectures

Three types of architectures form the core pillars of contemporary enterprises: social, physical, and digital. Each oversees a distinct yet interconnected aspect of the enterprise's existence.

Social architecture provides an insightful understanding and orchestration of social dynamics, forming the backbone of any organization. Traditional physical architecture, on the other hand, constructs the physical infrastructure where enterprises operate and flourish. Concurrently, digital architecture shapes the computational networks that fuel the enterprise, a structure pulsating with data and insights.

The enterprise's social architecture establishes legitimacy through the cultivation of accountability, trust, cultural coherence, and organizational norms. In contrast, the physical and digital architectures each manifest unique yet interconnected dimensions of the enterprise's identity and operational capability.

The physical architecture of an enterprise offers tangibility and presence. It is through this physical boundary that the enterprise asserts its material existence, providing a concrete space for interaction and collaboration. The design and location of its physical spaces communicate the enterprise's values and priorities, offering a sense of permanence and stability.

CHAPTER 1 PROBLEM

Conversely, the digital architecture embodies agility and connectivity. In today's interconnected world, an enterprise's digital boundary defines its adaptability and responsiveness. It facilitates real-time communication and data exchange across global networks, enabling the enterprise to maintain a dynamic presence in virtual spaces.

Together, these physical and digital boundaries, along with the social architecture, create a comprehensive framework within which the enterprise operates. Each aspect reinforces and complements the others, contributing to a robust, resilient, and responsive organization.

Achieving a harmonious alignment between the social, physical, and digital realms is more than an aspiration; it is a necessity. The key to this alignment lies in empowering individuals and collectives with physical and digital extensions, fostering an ecosystem where all elements can thrive symbiotically.

CHAPTER 1 PROBLEM

Figure 1-12. *It turns out that the convergence of the social, physical, and digital realms happens exclusively in Evolveburgh. Here, the grand prism of convergence casts a spectrum so intense that even the local pigeons have joined the debate on quantum physics, pondering whether Schrödinger's cat might secretly be a pigeon. Meanwhile, enterprise architects are wrestling with their spaghetti code of chaos, struggling to remember passwords to the digital labyrinth and embarking on a frantic quest for the mythical "Any" key, rumored to unlock all realms of possibility*

As enterprises navigate this complex interplay, the role of sociotechnical building blocks becomes paramount, guiding the enterprise toward a future that is not only efficient and responsive but also resonates with the core values of accountability and legitimacy.

In essence, the convergence of social, physical, and digital architectures poses a significant challenge. This intersection forms a complex matrix where each realm demands attention. Navigating this matrix requires more than just understanding each component; it necessitates a solution that can harmoniously integrate these diverse elements. This challenge highlights the need for innovative approaches

CHAPTER 1   PROBLEM

and solutions that can effectively manage this convergence, transforming it from a problem into an opportunity. The goal is to create an enterprise that is not only efficient and responsive but also embodies the core values of accountability and legitimacy. This is the pressing issue that enterprises face today, and solving it will pave the way for a future that is ready to embrace the unfolding potentials of a dynamically balanced enterprise.

The Information Masquerade: Does It Really Flow?

In contemporary enterprises, the flow of information and communication often follows complex and suboptimal patterns. These patterns, which include interconnected, intertwined, circular, fragmented, delayed, uncontrolled, and insecure flows, are influenced by several factors.

Complicated dependencies among business units often lead to convoluted and intertwined communication flows, escalating coordination costs. This is particularly true in cases of reciprocal interdependence, where the output of one unit cyclically becomes the input for another, significantly increasing the intensity of interaction and communication needs.

The complexity of technology also contributes to these challenges. Today's enterprises are often entangled in a complex network of interconnected platforms, databases, and applications. The sheer complexity of this technological landscape makes it challenging to fully comprehend and safely modify, contributing to the labyrinthine nature of information flows.

Misunderstanding the nature of work and inter-entity relations also play a role. Often, there is a lack of understanding regarding the division of responsibilities and how it necessitates coordination, which in turn relies on effective communication. This lack of clarity directly impacts the efficiency of communication flows.

CHAPTER 1   PROBLEM

Figure 1-13. *In Evolveburgh, modern problems require ancient solutions. In an effort to streamline communications, the government office has banned "Reply All" in favor of carrier pigeons. Turns out, "Fly All" was much more manageable! And on the plus side, the office's fertilizer expenses have never been lower*

Furthermore, there is frequently a lack of insight into the nature of relationships both within the enterprise and between its entities and external environments. Consequently, the same communication channels and protocols are repeatedly used for various relational dynamics, leading to inefficiencies and confusion.

A common issue is the use of identical communication channels and protocols for transmitting both data/information and digital products. This lack of differentiation can lead to inefficiencies and miscommunications, as the nuances of each type of transmission are not adequately considered.

CHAPTER 1   PROBLEM

Addressing these challenges requires a partitioning of the enterprise into a network of sufficiently autonomous units, which would not only streamline the structural organization but also harmonize the relations between units. It strikes a balance between maintaining necessary dependencies and reducing the needs for coordination and communication. This partitioning minimizes disruption and waste, leading to more effective and efficient information and communication flows.

By understanding and reorganizing the nature of work, inter-entity relations, and the content of communications, enterprises can significantly enhance their operational effectiveness and adaptability. This approach paves the way for enterprises that are not only more efficient and responsive but also better aligned with their core values of accountability and legitimacy.

Command and Control: Still Around After All These Years

In enterprise computing, coordination remains a relatively unexplored frontier. This important area fundamentally relies on communication, which, in turn, is built upon a foundation of robust connectivity. It serves as a critical mechanism guiding the interactions within various segments of the enterprise and orchestrating their engagement with elements in the broader environment.

Traditionally, the realm of business operations relied heavily on military-derived command-and-control doctrines, an approach gradually revealed to be incompatible with the nuanced complexities woven into enterprise operations and customer interactions. This centralized organization and coordination mechanism demands considerable resources in design, implementation, and deployment, transpiring as a process that's not just financially draining but also time-consuming.

CHAPTER 1   PROBLEM

Yet, despite its glaring inefficiencies and governance challenges, this antiquated method persists stubbornly, an emblem of an era that seemingly refuses to bow out gracefully.

Figure 1-14. *In Evolveburgh, "Command-and-Control" has morphed into "Post-it-and-Patience," turning the office into a modern art installation of sticky notes. This transformation serves as a daily reminder that work, much like a Post-it, is temporary, colorful, and tends to stick to you when you least expect it. The latest management upgrade, "Chaos-and-Mirror," has employees reflecting amidst the chaos, revealing that work, much like a mirror, can be enlightening, perplexing, and sometimes make a lead janitor look like a CEO. But hey, in the mirror of chaos, everyone gets a promotion!*

However, amidst the complexities of rigid control mechanisms, a new paradigm emerges, offering an alternative to cumbersome command-and-control processes: stigmergy. This is a time-tested coordination technique that nature, particularly among social insects, has refined over millions of years. Stigmergy demystifies the concept of collective action, where the result of one agent's work within the environment triggers subsequent activities, either by the same agent or others. This environmental

CHAPTER 1 PROBLEM

mediation ensures tasks are completed in the correct sequence, eliminating the need for detailed planning, direct interaction, or control. Thus, it provides a solution to the complex coordination paradox.

Figure 1-15. *Evolveburgh's mayor, in a "buzz" of brilliance, dispatches his council to learn the stigmergic behavior of social insects from the forest... completely missing the memo that a squadron of bees had been running a "Pollination 101" masterclass right in the flower bed under his office window*

CHAPTER 1   PROBLEM

Bridging the Divide: From Direct to Indirect Coordination

Two prominent coordination mechanisms—the traditional command-and-control (direct coordination) and stigmergy (indirect coordination)—each offer unique approaches to managing tasks and workflows within an organization. While command-and-control relies on explicit instructions and hierarchical decision-making, stigmergy allows for self-organization and spontaneous order, with actions triggered by changes in the environment. Both mechanisms have their strengths and can be effectively utilized depending on the specific needs and context of the enterprise. Understanding and balancing these mechanisms can lead to more efficient and adaptable operations, ultimately contributing to the overall success of the organization.

Despite these potential benefits, direct coordination, whether executed directly by human actors, such as managers or administrators, or facilitated by Business Process Management (BPM) software, presents its own set of challenges often falling short of fulfilling its promises due to a multitude of factors.

Firstly, there is often a lack of understanding and difficulty in discovering implicit processes. These hidden or non-obvious processes can be critical to the functioning of an enterprise, but their elusive nature makes them difficult to manage effectively with traditional BPM tools.

Secondly, excessive standardization can stifle innovation and flexibility. While standardization can bring about efficiency and consistency, an overemphasis can lead to a one-size-fits-all approach that fails to accommodate the unique needs and contexts of different processes and tasks.

Thirdly, the high costs associated with the initial design of BPM processes can be prohibitive. This, coupled with a general lack of BPM skills within organizations, can hinder the effective implementation and utilization of BPM software.

CHAPTER 1   PROBLEM

Furthermore, the lack of flexibility and adaptability of BPM services, the high costs of BPM software, and the automation of inefficient processes all contribute to the underperformance of direct coordination mechanisms. Additionally, resistance to change, often due to a lack of understanding or fear of the unknown, can further impede the successful implementation of BPM initiatives.

Given these challenges, a paradigm shift toward indirect coordination is required. This approach enables the dynamic discovery and formalization of business processes, which are likely to resemble networks of events rather than rigid sequences of tasks as in current BPM models.

This shift mirrors the success of social insects and their approach to work based on indirect coordination and communication. Just as ants leave pheromone trails for others to follow, leading to efficient food foraging without centralized control, enterprises can benefit from similar principles. By leaving "digital pheromones" in the form of data and metadata and visualizing the respective emerging coordination structures, enterprises can coordinate their activities organically, leading to more efficient and adaptive processes.

Learning from nature and embracing the power of stigmergy will help enterprises to transform their operations from a rigid, top-down command-and-control model to a more flexible, bottom-up, self-organizing system, paving the way for increased efficiency, adaptability, and resilience in the face of change.

Email: An Unintended Tool for Stigmergic Coordination

However, the shift toward stigmergic coordination encounters a significant obstacle: the lack of an effective indirect coordination system offered by the software industry. This void has led to the unexpected adoption of email as a primary tool for stigmergic coordination—a trend that underscores the dominant role of email in data and information transfers within organizations. Yet, this usage reveals a distinct challenge—the incapacity of general technologies like email to create explicit coordination artifacts, a deficiency that limits the potential of stigmergic coordination. Nevertheless, this challenge presents the software industry with two immediate opportunities: the chance to add a coordination layer to email and to develop next-generation BPM systems centered on indirect coordination.

Figure 1-16. Back from a "buggy" sabbatical in the forest, Evolveburgh's council members champion email as their new "digital ecosystem" for stigmergic coordination. Citing fewer mosquito bites and superior WiFi as key benefits, they quip that the ants were too "ant-iquated" for tech support. And the mayor? Well, in the world of stigmergy, he's as necessary as a beekeeper in an ant colony!

CHAPTER 1   PROBLEM

To fully leverage the transformative potential of stigmergy—a mechanism that has driven the evolutionary success of social insects—we must pave the way for directed emergence. This approach encourages the swift identification of efficient work processes and promotes seamless collaboration between computational and human agents. This strategy aims to reduce dependencies and streamline information flows, advocating for a flexible, adaptable logical structure that can readily adapt to future advancements with agility and ease.

As we approach the threshold of a significant revolution, we foresee a substantial shift in coordination dynamics, similar to the transformative wave brought about by the Internet. The impending Coordination Revolution promises to reshape our comprehension of enterprise mechanisms, potentially sooner than anticipated, guiding us toward a future where coordination transcends being merely a functional necessity and evolves into a mastered art form.

To Transform or Not to Transform: The Urgency of Choice

Enterprises, similar to living organisms, owe their survival to the cyclical nature of change and renewal that permeates their structure. The goal is to build enterprises that are resilient to the erosive effects of time, especially in the digital realm where time's impact is swift and profound. To withstand these forces, enterprises must facilitate both macro- and micro-level changes, dexterously evolving to meet user requirements while minimizing disruptions and potential ripple effects. However, as time progresses, larger interventions become less flexible and less effective, underscoring the need for adaptive strategies and frameworks.

In today's digital age, enterprises cannot afford to be static; they must evolve into sophisticatedly designed architectural entities that reflect ongoing technological and societal transformations. Just as a living organism continually renews its cells, an enterprise thrives on constant change and renewal, embodying a dynamic structure that continuously adapts and transforms. The enterprise's longevity is a testament to its ability to metamorphose and subtly rejuvenate its components to withstand the test of time, even in the face of an accelerated digital timeline.

Figure 1-17. *"To invoice or not to invoice, that is the Deja Vu," chant the high-priced consultants hired for Evolveburgh's quirky transformation, which inadvertently created the "Deja Vu" department. Here, it's Groundhog Day every single day. Executives and employees excel in predictability, mastering the art of performing the exact same tasks with impressively fresh confusion. Their motto? "We may not remember yesterday, but our bank account sure does!"*

Key Takeaways

This chapter has explored the pivotal crossroads at which enterprise architecture currently stands, amidst rapid technological evolution and shifting market dynamics. Here are the essential insights and imperatives for transforming enterprise architecture:

CHAPTER 1   PROBLEM

- **Need for a cohesive grand vision**: The enterprise often lacks a cohesive grand vision, having historically been shaped more by intuition and happenstance than by deliberate strategic design. This gap underlines the urgency for a comprehensive, strategic approach that integrates all aspects of enterprise architecture.

- **Limitations of reductionist approaches**: Traditional reductionist approaches, while providing analytical rigor, fail to capture the holistic essence of organizational architecture. These approaches often prioritize individual disciplines or systems at the expense of overall architectural cohesion, underscoring the need for a more integrative approach.

- **Conflict among stakeholders**: Competing priorities among enterprise stakeholders often lead to conflicts, particularly in complex scenarios such as mergers and acquisitions. The absence of enterprise architects in critical decision-making processes highlights the need for their proactive involvement to balance business interests and architectural integrity.

- **Architectural opacity**: Architects often struggle to fully comprehend what an enterprise looks like due to its scale and complexity. This complexity necessitates a foundational template or blueprint that provides a clear, adaptable starting point for understanding and shaping any enterprise.

- **Harmonious alignment of architectures**: Achieving alignment between social, physical, and digital architectures and empowering individuals and collectives with the necessary tools and frameworks supports a symbiotic ecosystem where all elements can thrive.

- **Optimization of information flow**: In many enterprises, information and communication flow in complex and suboptimal patterns. A reorganization of the enterprise into sufficiently autonomous units could streamline structural organization and improve communication efficiency.

- **Shift toward indirect coordination**: The limitations of direct coordination, whether human-led or facilitated by BPM software, demand a shift toward indirect coordination. This new paradigm promotes the dynamic discovery and formalization of business processes, which are more likely to resemble networks of events than rigid task sequences.

- **Cyclical nature of enterprise evolution**: Just as living organisms adapt and renew themselves, enterprises must embrace change and renewal to remain resilient. The goal is to build transformable enterprises that are not just resistant to time's erosive effects but are also proactive in continuous adaptation and transformation.

Each of these takeaways not only highlights a specific challenge within the field of enterprise architecture but also proposes a direction for future development and innovation. These insights pave the way for a more responsive, integrated, and dynamic approach to enterprise architecture, ensuring that organizations can thrive in an increasingly complex and unpredictable global landscape.

CHAPTER 2

Idea

> *Embarking on a "Why-dunit" with a compass set to South, the author delicately strips away the mystique of* Unit Oriented Architecture: Building Large Sociotechnical Systems, *proving that even in the most serious disciplines, we can still find time to read from right to left.*

As we move forward from our exploration of the prevailing issues within enterprise architecture, it is clear that traditional approaches are increasingly inadequate for today's dynamic and complex enterprise environments.

As John Zachman, a renowned thought leader in enterprise architecture, points out (Zachman 2015a):

> *The Enterprises of today ... have never been engineered. They happen ... incrementally ... over the life of the Enterprise as it grows and requires formalisms ... "systems," manual and/or automated.*

This piecemeal evolution results in architectures that lack cohesive planning and integration. Zachman further elaborates:

> *IT has been manufacturing the Enterprise (building systems) for 70 years or so... but the Enterprise was never engineered. Therefore, IT has not been manufacturing the Enterprise... they have been manufacturing PARTS of the Enterprise ... and the parts don't fit together (they are not "integrated").*

CHAPTER 2   IDEA

These insights underscore the need for a paradigm shift from traditional approaches to more holistic, integrated methodologies that address the complexities and interdependencies of modern enterprises. Chapter 1 highlighted a myriad of challenges—ranging from the lack of a cohesive grand vision and the pitfalls of reductionist methods to the conflicts among competing stakeholders and the sheer opacity that clouds our understanding of an enterprise's true structure. These challenges underscore the urgent need for a paradigm shift in how we conceptualize and implement enterprise architecture.

Transitioning into the next chapter, we delve into the foundational ideas that propel us toward a novel framework for architecting enterprises. This chapter, titled "Idea," introduces the core concepts that underpin our proposed shift away from traditional architectural methods. It is an exploration aimed at redefining the scope and application of enterprise architecture to better suit the vast, sophisticated, and inherently interconnected world we operate within.

Foundations for Change: The Imperative for a New Enterprise Architecture Paradigm

Recognizing the monumental task of selling the idea of a complete overhaul of the modern approach to enterprise architecture, this book proposes a radical but necessary shift.

CHAPTER 2 IDEA

Figure 2-1. *In Evolveburgh, the citizens' enthusiasm for seismic changes finally shook things up too much, prompting the tectonic plates to serve a restraining order—clearly, even nature has its limits!*

Instead of narrowly focusing on the issues currently overlooked by both the industry and practitioners, the book embraces a broader and more expansive perspective: the architecture of large sociotechnical systems. This shift broadens the scope beyond conventional enterprise architecture frameworks to a universal design methodology that is applicable to a wide array of complex sociotechnical systems.

The necessity for this new approach stems from several pressing needs within the realm of enterprise architecture. First, there is a clear demand for an architecture style that is comprehensible and universally understood by all stakeholders involved in or affected by enterprise design (development) and operations. Traditional architectures often result in knowledge silos and communication barriers that hinder effective decision-making and operational efficiency. By developing a universally understandable architecture, we ensure that stakeholders from diverse domains—ranging from IT professionals to business executives and from end users to regulatory bodies—can effectively collaborate and contribute to the system's success.

Second, the dynamic nature of modern business environments requires architectures that support transformations at any scale—from incremental adjustments to major overhauls—ensuring adaptability and responsiveness across all levels of the enterprise. Enterprises must be able to pivot quickly in response to market changes, technological advancements, or regulatory updates. This book proposes an architecture that facilitates easier and more predictable transformations, thereby enabling enterprises to evolve continuously and responsively.

Third, recent advancements in technologies, particularly in artificial intelligence, monolithic enterprise data platforms (see "How to Move Beyond a Monolithic Data Lake to a Distributed Data Mesh" by Zhamak Dehghani), and monumental, hard-to-decompose business solutions like Salesforce, have raised new concerns about the accountability and transparency of enterprise systems. The proposed unit oriented architecture addresses these concerns by defining clear, manageable building blocks (units) that include mechanisms for tracking their performance, interactions, and evolution. Each unit's design encapsulates its functionality and governance, thereby enhancing accountability and traceability across the system.

Finally, streamlined communication flows are vital for the effective operation of large sociotechnical systems. The complexity of these systems often leads to fragmented communication channels that can obscure critical information and slow response times. A unit oriented enterprise architecture simplifies these communication paths by clearly delineating interfaces and interaction protocols between units. This not only improves information exchange but also enhances system-wide coherence and coordination.

In essence, this book advocates for a new architectural paradigm that addresses the shortcomings of traditional frameworks through a comprehensive, adaptable, accountable, and efficiently communicable approach. By doing so, it aims to equip enterprises with the resilience and agility needed to thrive in the rapidly changing modern landscape.

Having established the need for a transformative shift in our approach to enterprise architecture, we now turn to a detailed exploration of the underlying principles that drive this new paradigm. Each of the next sections delves into a critical aspect of our proposed framework, systematically addressing the essential questions: Why Unit Orientation, Why Architecture, Why Large, Why Sociotechnical, and Why Systems. These discussions aim to explain the components of the *Unit Oriented Architecture: Constructing Large Sociotechnical Systems in the Age of AI* book title in reverse order, providing a comprehensive understanding of how each element supports the overarching framework. By examining these foundational elements, we can better appreciate the full scope of the proposed changes and their potential to reshape enterprise architecture into a more effective tool for managing today's complex sociotechnical systems.

Why Systems

In the context of *Unit Oriented Architecture: Constructing Large Sociotechnical Systems in the Age of AI*, the term "system" is pivotal and is used within the context of systems thinking. This discipline centers around a comprehensive understanding of the complex interdependencies that characterize the entirety of an entity rather than isolating its parts. Systems thinking promotes an examination of the whole architecture—comprising interconnected parts—which collectively influence the overall functionality and behavior of the system. This methodology is invaluable for navigating and managing complex environments, where diverse human, technological, and organizational factors interplay dynamically.

Defining a "System" in This Framework

Within this framework, a "system" is conceptualized as an entity comprising uniformly structured interconnected components that interact among themselves and with entities in their surrounding environments, collectively enhancing the system's overall success.

CHAPTER 2 IDEA

Figure 2-2. *In Evolveburgh, there's a by-law that requires citizens to think about systems and systems to think about citizens. It's gotten to the point where even the coffee machines are having existential crises and the traffic lights are contemplating the meaning of red, yellow, and green*

While the system's inherent interconnectedness and interdependence remain critical attributes, the unit oriented enterprise architecture distinctively strives to maximize the autonomy of each unit. This autonomy reduces dependencies, thereby diminishing the need for extensive coordination and communication. Additionally, this approach aims to minimize the impact of any single component's actions or changes on others and the system as a whole. However, it still underscores the importance of a holistic view in the design and management of these systems, ensuring that even with enhanced autonomy, the overarching system's integrity and functionality are preserved.

CHAPTER 2   IDEA

Systems Thinking in Enterprise Architecture

Systems thinking offers many benefits in enterprise architecture. By adopting this holistic view, organizations can achieve more effective problem-solving and foster innovation. It allows architects to understand and address the complexities of the enterprise environment, taking into account the interrelationships and interdependencies between components of the system.

In the context of enterprise architecture, systems thinking involves viewing the enterprise as a cohesive whole, rather than a collection of individual parts. This perspective enables architects to identify patterns and structures that may not be apparent when looking at components in isolation. It promotes a broader understanding of the enterprise, which can lead to more informed decision-making and strategic planning.

Systems thinking also fosters a culture of collaboration and communication within the organization. It breaks down silos and encourages cross-functional teams to work together toward common goals. This collaborative approach can lead to more innovative solutions and improved organizational performance.

It also encourages a more integrative approach to organizational and technological developments, ensuring that changes in technology align with the human and technological elements of the enterprise, thereby enhancing overall efficiency and adaptability. By considering the impact of technological changes on people and processes, organizations can ensure a smoother transition and minimize disruption.

Providing a comprehensive framework for understanding and managing the complexities of the enterprise, which leads to more effective problem-solving, increased innovation, and improved organizational performance, systems thinking serves a powerful tool for enterprise architecture.

CHAPTER 2   IDEA

Essential Characteristics of Systems

The characteristics of systems within the discipline of systems thinking can be divided into structural and nonstructural categories. Each category plays an important role in understanding how systems function and interact both internally and with their environment.

Structural Characteristics

- **Holism**: Systems are perceived as cohesive wholes rather than mere aggregations of parts. This holistic view is essential because the behavior of the system cannot be fully predicted or understood by analyzing its individual components in isolation. The system's overall behavior emerges from the complex interactions among its parts.

- **Organization and self-organization**: Systems may exhibit organization, where components are deliberately arranged to achieve specific goals, or self-organization, where components naturally find an efficient arrangement without explicit external guidance. This self-organizing behavior is particularly evident in adaptive and resilient systems.

- **Boundaries**: Every system has defined boundaries that delineate it from other systems and the external environment. These boundaries can be social, physical, digital, or conceptual, defining the scope and scale of system operations.

- **Configurability**: Systems are designed with varying degrees of configurability, allowing them to be adapted to different needs or conditions. This adaptability is critical for systems to remain relevant and functional as external conditions change.

- **Hierarchy**: Systems are often organized hierarchically, with smaller systems nested within larger systems. Each level operates as a system in its own right, with its subsystems and supersystems, contributing to a complex, multilayered architecture.

- **Emergence**: Systems exhibit emergent properties that are not present in the individual components alone. These properties arise from the interactions between components and often represent new abilities or behaviors that are integral for the system's functionality.

- **Interdependence**: The components within a system are interdependent, meaning that changes in one part can influence other parts. This interrelation emphasizes the need for a coordinated approach in system management and design.

Nonstructural Characteristics

- **Feedback loops**: Many systems incorporate feedback loops, where outputs of the system are recirculated back as inputs. This process influences subsequent system behavior, enhancing its ability to self-regulate and adapt to new conditions.

- **Dynamics**: Systems are inherently dynamic; they change and evolve over time. Their ability to adapt, evolve, and self-organize is essential for long-term survival and effectiveness, especially in rapidly changing environments.

- **Nonlinearity**: The behavior of systems often shows nonlinearity, where small inputs or changes can produce disproportionately large effects, and conversely, large inputs may have minimal impact. Understanding these nonlinear dynamics is vital for effective system management and prediction of system behavior.

- **Exchange of information, matter, and energy**: Systems engage in continuous exchange of information, matter, or energy with their components and the environment. This exchange is critical for system operations and can influence system stability and performance.

- **Learning capabilities**: Advanced systems, particularly those involving technological and human components, often possess learning capabilities that allow them to improve performance over time based on past experiences. This characteristic is increasingly important in adaptive systems that must cope with variable and often unpredictable environments.

These essential characteristics outline the complexity and dynamism of systems in the context of unit oriented architecture. Understanding these traits helps in designing more efficient, adaptable, and resilient systems, capable of thriving in complex sociotechnical landscapes.

Integrated Systems and Modular Systems

In systems design, it is common to categorize systems into two broad types: integrated systems and modular systems. This classification offers different approaches to the organization and functionality of components within the system, each with its unique strengths and applications. The choice between these types depends largely on the specific requirements of the project or organization, as well as the desired flexibility, scalability, and maintainability of the system.

Integrated Systems

Integrated systems are characterized by their tightly coupled components that work together as a cohesive whole toward a unified goal. In these systems, each component is specifically designed to function in harmony with the others, often resulting in optimized performance and efficiency. However, this tight integration can also introduce challenges, particularly when it comes to modifying or scaling the system. Changes to one part of an integrated system may necessitate significant adjustments or redesigns throughout, which can be time-consuming and costly.

Modular Systems

In contrast, modular systems consist of loosely coupled, interchangeable modules that maintain their functionality independently but can be combined to provide complete system functionality. This design allows for greater flexibility and scalability, as individual modules can be added, removed, or replaced without impacting the overall system. Such systems are particularly well suited to environments where needs and conditions change frequently, allowing organizations to adapt without extensive overhauls. A practical example of a modular system is a content management system (CMS), where various plug-ins or modules—such as those for SEO, social media integration, or analytics—can be activated or deactivated according to user needs and preferences.

CHAPTER 2   IDEA

Choosing Between Integrated and Modular Systems

The decision to design a system as either integrated or modular should be based on the operational context and specific needs of the organization. Integrated systems are often favored when there is a need for high performance and efficiency with stable, well-defined requirements. On the other hand, modular systems are advantageous in dynamic environments where the ability to quickly adapt to new technologies or changing business conditions prove essential.

Both integrated and modular systems offer valuable solutions in the design of large sociotechnical systems, each aligning with different strategic objectives and operational contexts. The choice between these systems—or the decision to blend elements of both into a hybrid system—depends on the balance of flexibility, scalability, and efficiency needed to meet the challenges of building and maintaining complex systems in various environments.

Alternatives to Systems

While systems thinking provides a comprehensive framework for understanding complex entities, other perspectives can also be considered, each with its unique applications and limitations:

- **Collections of independent components**: This approach treats components as stand-alone elements without interdependencies. For instance, consider a software suite comprising independent applications such as word processors, spreadsheets, and presentation tools. Each application operates independently, without the need for interaction with others. While this can simplify development and reduce complexities associated with interconnected systems, it may also limit the overall efficiency or adaptability of the system since synergies and potential enhancements from component interaction are overlooked.

- **Organic structures**: Organic structures are similar to biological organisms in that they evolve naturally or adaptively over time. This design philosophy eschews systematic, predefined pathways in favor of evolution and adaptation to meet emerging needs and respond to environmental changes. While this allows for flexibility and spontaneity in development, it may lack the predictability and consistency offered by more structured, systematic planning.

While there are alternatives to systems, none provide the holistic benefits essential for managing the complexity of large integrated sociotechnical systems. Systems not only facilitate a deeper understanding of such complexities but also enhance the capability to design architectures that are robust, adaptable, and cohesive.

Why Sociotechnical

Sociotechnical systems are comprehensive structures that seamlessly integrate the social, physical, and digital dimensions of human capabilities. These systems are not simply a theoretical concept but a practical necessity that underpins much of our modern life, including areas as diverse as healthcare, education, transportation, and communication. The term "sociotechnical" refers to the symbiosis between humans, social systems (people, culture, societal norms), and technological systems (tools, processes, technologies), highlighting that the effectiveness of any significant system hinges on this integrated approach.

CHAPTER 2 IDEA

The crux of sociotechnical systems lies in their capacity to support and enhance the three foundational human abilities: the ability to make (physical), the ability to compute (digital), and the ability to relate (social). These abilities are not just critical milestones of human development; they are the dynamic forces that drive our evolution and the evolution of the systems we create. When designing and operating any significant human-made systems, integrating these three realms is paramount. This integration ensures that our systems are not merely effective and efficient but are also equitable and adaptable, aligning with the ever-evolving human landscape.

Most complex systems constructed to support human activities must inherently be sociotechnical in nature. By ensuring that systems fundamentally support these mutually exclusive and collectively exhaustive human capabilities, we position them to better meet the challenges of a complex world. Understanding and reinforcing the interplay among the physical, digital, and social dimensions is essential for creating systems that are capable of adapting and thriving amidst continuous change. In doing so, we commit to a design philosophy that views effective system design not just as a technical challenge but as a comprehensive approach that respects and utilizes the strengths of human interaction and technological innovation.

Figure 2-3. *Evolveburgh's greatest minds were so concerned with AI going rogue, they made them sociotechnical. Now, each AI is surrounded by a human chain of hand-holding techies, ready to jump into action at the slightest hint of a glitch. It's like a never-ending game of "Red Rover" with the AI in the middle. The motto? "We may not stop the AI from taking over, but we'll sure give it a good group hug first!"*

CHAPTER 2 IDEA

The Three Foundational Capabilities
The Ability to Make (Physical)

The ability to make—referring to the physical capability of humans to create, build, and manipulate the material environment—is a fundamental human attribute that has long been a cornerstone of societal advancement.

Figure 2-4. In Evolveburgh, the ability to make is so celebrated that they've started hosting annual "Make-a-thon" competitions. One year, a contestant took it literally and made an actual "thon"! Now, he's the proud owner of a lifetime supply of duct tape, because in Evolveburgh, they believe, "If you make a thon, you better be ready to give it a duct tape makeover!"

CHAPTER 2   IDEA

From the crafting of primitive tools to the construction of skyscrapers, this capability encapsulates the tangible aspects of human ingenuity and engineering. In the context of sociotechnical systems, the physical capability plays a critical role in shaping the environments in which these systems operate and the interfaces through which humans interact with technology.

The physical capabilities of humans encompass a broad spectrum of activities essential to sociotechnical systems, extending from the creation of simple tools to the development of sophisticated manufacturing processes and infrastructure systems. These capabilities enable us not only to transform raw materials into useful objects and structures but also to process, transport, store, maintain, and recycle materials and products in ways that optimize efficiency and sustainability. The significance of physical making thus extends beyond mere construction; it involves a deep understanding of material properties, engineering principles, and ergonomics, all indispensable for creating efficient, durable, and user-friendly physical systems. These capabilities collectively enhance our ability to control and manipulate our environment, ensuring that physical structures and systems can support the dynamic needs of both digital and social elements within a sociotechnical framework.

Within sociotechnical frameworks, the physical capability is linked to both the digital and social realms. Physical systems provide the infrastructure necessary for the deployment of digital technologies and the embodiment of social interactions. For example, in smart cities, physical components such as sensors, networks, and built environments are designed to interact seamlessly with digital systems that manage data and automate processes, all while considering the social dynamics of the urban populations they serve.

The design of physical systems within sociotechnical contexts also reflects a deep understanding of human physical interaction and ergonomics, ensuring that technologies are accessible and practical. Considerations such as the placement of physical controls, the design

57

CHAPTER 2　IDEA

of public spaces, and the accessibility of buildings and transportation systems all illustrate how physical making is influenced by and influences sociotechnical systems.

The integration of physical capabilities in enterprises is not just about functionality but also about sustainability and resilience. The physical structures we create must not only support the immediate needs of digital and social entities but also be adaptable to future changes and challenges. This requires a forward-thinking approach to materials science, architectural design, and urban planning, ensuring that physical systems can evolve alongside their digital and social counterparts.

In summary, the ability to make in the physical realm is not just a testament to human creativity and engineering prowess but a fundamental aspect of sociotechnical systems that ensures these systems are grounded in the practical realities of the physical world. By effectively integrating this capability, we enhance the utility, durability, and adaptability of sociotechnical systems, making them more capable of meeting the complex demands of modern societies.

The Ability to Compute (Digital)

The ability to compute encompasses the digital capabilities critical to modern enterprises, extending well beyond basic computational tasks to include the processing, transporting, storing, maintaining, and recycling of digital representations. This suite of capabilities enables sophisticated data analytics, machine learning, artificial intelligence, and expansive communication networks, all integral to transforming raw data into actionable insights that drive decision-making and innovation across societal sectors.

Digital capabilities allow us to efficiently handle vast quantities of data, ensuring that information can be processed, stored, and retrieved with agility and precision. This not only enhances decision-making processes but also secures the longevity and accessibility of information, which are paramount for ongoing operations and future developments. Digital tools and platforms facilitate the movement of information across vast distances at nearly instantaneous speeds, supporting global communication and collaboration.

Figure 2-5. In Evolveburgh, computing transcends number crunching to become a life journey. From data processing to recycling, the city even holds funerals for retired data. The most touching eulogy? "10110101 was just a byte, but it left a terabyte of memories." And the preferred epitaph? "Rest in Pixels."

In sociotechnical systems, digital capabilities integrate deeply with physical and social components, providing a critical infrastructure that supports dynamic interactions and efficient system operation. Processing information through digital means allows for real-time analysis and response, essential in environments such as healthcare, where patient monitoring systems provide immediate data to medical staff, or in financial services, where transactional data must be processed quickly and securely.

The transport of information enabled by digital technologies breaks down geographical and organizational barriers, enhancing the capacity for remote collaboration and operation. Storing information digitally not only facilitates efficient data management but also ensures that historical data is preserved for regulatory compliance and future reference. Maintaining digital systems involves regular updates and security measures to safeguard against threats, ensuring system integrity and reliability over time.

Moreover, the ability to recycle digital resources—updating software, repurposing data, and responsibly managing electronic waste—reflects a commitment to sustainability and adaptability in system design. This not only extends the life cycle of digital assets but also reduces the environmental impact of technological advancement.

To summarize, the ability to compute in the digital realm is a foundational element of sociotechnical systems, ensuring that these systems are adaptable, scalable, and capable of evolving with the changing needs of society. Digital capabilities are instrumental in bridging the gap between the static infrastructure of the physical world and the dynamic requirements of social systems, making them essential for the continued relevance and efficiency of sociotechnical frameworks.

The Ability to Relate (Social)

The ability to relate is a comprehensive term that covers the entire social domain. It includes fundamental social capabilities such as associating, connecting, communicating, empathizing, sympathizing, understanding, collaborating, influencing, and adapting. Essentially, it encapsulates all aspects of social interaction and connection and more. This suite of capabilities forms the foundation for human interaction and governs how individuals and groups coordinate and coalesce within various systems. These capabilities are not only about facilitating communication but also about deepening understanding and fostering genuine connections that enhance collaborative efforts and collective resilience.

Figure 2-6. In Evolveburgh, social skills go beyond delivering the classic "404 error" joke—"Error 404: Bathroom not found!"— at the annual "404 Festival." It's also about "minimizing" those cringeworthy moments when you accidentally call your boss "Mom" and "defragmenting" your brain after endless meetings. It's about learning when to "log out" of conversations going nowhere and how to "clear your cache" of those not-so-glorious office party gaffes. And remember, always keep a "backup" plan for when your neurons break from reality to question their existence and essentially go on strike. Because in Evolveburgh, life's glitches always come with a chance to hit refresh!

Social capabilities extend beyond the mechanics of simple interaction, reaching into the realms of effective, meaningful communication and the nuanced understanding of human emotions and social dynamics.

CHAPTER 2   IDEA

They are essential for building robust networks and relationships, which are the bedrock of collaborative and community-focused endeavors. In enterprises, these social interactions are pivotal in ensuring that technology aligns with human values and societal needs, promoting inclusivity and accessibility.

In the context of sociotechnical systems, the ability to relate is central to both the design and operation of these complex assemblies. Effective systems leverage social capabilities to ensure that the technology serves its intended human function, adapting to and anticipating the needs of diverse user bases.

Through its multifaceted components, the ability to relate plays a fundamental role in handling the key operational areas of sociotechnical systems. These social capabilities are essential not only for effective communication but also for directing, guiding, and coordinating the various functions and processes within the system. By effectively harnessing these social interactions, sociotechnical systems can ensure that operations are seamlessly aligned with both the immediate needs and the broader goals of the system. Furthermore, these capabilities help to constrain and adapt operations in response to dynamic environmental and societal changes, ensuring that the system remains responsive and relevant. This holistic integration of social capabilities ensures that the system not only operates efficiently but also evolves in harmony with the human elements it is designed to serve.

Moreover, the social capability to connect and associate within and across communities facilitates the expansion of networks and the integration of broader societal insights, enriching the system's capacity to innovate and evolve. Systems that embody these principles are more likely to succeed in creating supportive, adaptive environments that respect and respond to the human elements at their core.

In sum, the ability to relate is a foundational element of sociotechnical systems, playing a critical role in ensuring that these systems are not only technically efficient but also socially attuned and responsive.

By prioritizing social capabilities, system designers can create more inclusive, effective, and adaptable environments that truly meet the needs of their users and reflect the societal values they aim to uphold.

Integration of Capabilities in Sociotechnical Systems

Effective design of sociotechnical systems demands a nuanced understanding of the interplay between the three foundational capabilities: physical, digital, and social. The integration of these capabilities is not merely about ensuring each operates effectively on its own, but rather about harmonizing them in a manner that the system as a whole functions optimally, equitably, and sustainably. This approach ensures that sociotechnical systems are not only efficient in their operations but also considerate of human and environmental impacts, promoting long-term sustainability.

A well-integrated sociotechnical system considers the physical elements (the ability to make), the digital elements (the ability to compute), and the social elements (the ability to relate) as interconnected parts of a larger whole. Each capability must be tuned not only to support the functionality of the others but also to enhance the overall system's performance and adaptability. For instance, the physical infrastructure should support the optimal use of digital tools, which in turn should facilitate effective social interactions and organizational processes. Similarly, the digital infrastructure must be designed to enhance physical operations and enrich social engagements.

The integration also ensures that systems are equitable, providing all users with fair access to the system's resources and benefits. It also involves designing for diversity, ensuring that systems meet varied needs and are adaptable to different cultural contexts. Sustainability in this integration refers to the development of systems that maintain their functionality and relevance over time, without exhausting the natural or social resources they rely on.

CHAPTER 2   IDEA

An illustrative example of integrating these capabilities, albeit in a rudimentary form by modern standards, can be seen in the Seven Wonders of the Ancient World. These structures, marvels of ancient engineering, showcased an extraordinary integration of physical, digital, and social aspects.

Figure 2-7. *A group of ambitious citizens in Evolveburgh signed a petition to permanently move the Seven Wonders of the Ancient World to their city, only to gasp at the Hanging Gardens' water bill—higher than Mount Everest—and the fine print revealing no gardener was included, not even a part-time one! So, they turned to their own marvel: the city's self-proclaimed "Eighth Wonder," a triangle-shaped roundabout named Bermuda. This geometric oddity mysteriously dissolves political disputes. Who needs the Hanging Gardens when a roundabout can spin everyone back into harmony?*

Physical: The sheer size and scale of constructions like the Great Pyramid of Giza or the Statue of Zeus at Olympia demonstrated advanced capabilities in handling physical materials and constructing large-scale structures that have stood the test of time.

Digital: Although the digital in the modern sense did not exist, the designers and builders utilized advanced mathematical, architectural, and engineering principles. These can be thought of as the manipulation of mental representations of digital data, involving complex calculations and designs that ensured the feasibility and stability of these structures.

Social: The organization required to construct such wonders involved exceptional direction, governance, and coordination. The ability to mobilize and manage large groups of workers, the distribution of resources, and the alignment of these projects with cultural and political objectives illustrated sophisticated social organization.

These ancient wonders were not only technical feats but also served social and cultural purposes, symbolizing power, religious devotion, and technological prowess, thereby integrating social skills, norms, and values with making and computing capabilities.

In essence, the effective design of sociotechnical systems requires a balanced integration of physical, digital, and social capabilities. This integration must be directed toward creating systems that are not only operationally efficient but also socially equitable and environmentally sustainable. The historical example of the Seven Wonders provides a conceptual blueprint of how these integrations, even in ancient times, solved complex problems by combining diverse capabilities in a harmonized manner. Modern sociotechnical systems can draw valuable lessons from such integrations, aiming for a synergy that supports advanced technological societies today.

Why Large

In the design of sociotechnical systems, flexibility and scalability are paramount. While the principles discussed in this book apply to organizations of all sizes—from small startups to medium-sized enterprises—the approach is particularly adept at scaling to accommodate the needs of large and complex entities. This adaptability is indispensable in a world where organizations are increasingly operating on a global scale, with multifaceted interactions spanning multiple geographical and cultural boundaries.

Scalability of Sociotechnical Systems

The Unit Oriented Enterprise Architecture outlined in this book is designed to scale out to any size, making it an ideal framework for architecting vast, global enterprises. This scalability is not merely about growing bigger but about maintaining or even increasing efficiency and effectiveness as the system expands. Large organizations, particularly those with a geographical divisional structure, require a robust architecture that can handle the complexities associated with managing diverse units spread across the globe.

CHAPTER 2   IDEA

Prototype for Large Systems: The Global Enterprise

A global enterprise with a geographical divisional structure serves as a perfect prototype for architecting large sociotechnical systems within this framework. Such enterprises are characterized by their division into semi-autonomous geographic regions, each responsible for a segment of the organization's overall operations. This structure necessitates a design approach that allows for both autonomy and coordination across different levels and locations—a challenge adeptly met by the Unit Oriented Enterprise Architecture.

Inspired by a rich and flexible underlying metaphor that will be explored in greater detail later in this book, Unit Oriented Enterprise Architecture offers a sophisticated method for designing and managing large enterprises. The architecture's inherent flexibility allows for the customization of structures and processes to meet the unique demands of each unit within the organization while still aligning with overarching corporate goals and strategies.

CHAPTER 2   IDEA

Figure 2-8. In their quest for double XL-lence, Evolveburgh supersized all systems to "XXL." Once they welcomed hippos into their lives as exclusive pets, their world transformed into an unexpected hippotopia overnight. They developed a "hippotitude," indulged in hippo-doughnuts (essentially regular doughnuts but with hippo-sized holes), and their favorite catchphrase became, "What's hipponing?" They even started running hippothlons, which are like marathons, but with more wallowing and less running. Hypothetically speaking, the hippo-sized and hippotastic Evolveburgh is now on track to become the world's biggest tourist attraction. If you ever visit, don't get hippotized by their charm, just have a hipporiffic time! And remember, in Evolveburgh, "hippo" is more than just a word, it's a lifestyle!

Unit Oriented Enterprise Architecture supports the decentralization of decision-making, allowing local units to respond swiftly and effectively to changes in their specific market or regulatory environments. Yet, it also ensures that these units remain tightly integrated within the larger organizational framework, maintaining a cohesive strategy and unified operational standards.

CHAPTER 2   IDEA

To sum up, the emphasis on large systems within this book reflects the increasing need for architectures that can handle the complexities of large, dispersed, and diverse organizations. By adopting Unit Oriented Enterprise Architecture, enterprises can achieve the scalability required to operate effectively on a global scale, ensuring that they are equipped to handle the challenges of modern, dynamic business environments. This architecture not only supports growth and flexibility but also enhances the overall resilience and adaptability of the organization, making it capable of thriving in an ever-changing global landscape.

Why Architecture

When creating sociotechnical systems, the focus on architecture is critical. This book advocates for an architectural approach that concentrates on the purest elements of enterprise architecture: the components and their linkages. This method intentionally sets aside the aspects that have traditionally been entangled with enterprise architecture, such as aligning IT with business goals, defining strategy, optimizing processes, managing governance frameworks, overseeing technology roadmaps, and enforcing compliance. While these aspects are undeniably important, they can sometimes obscure the fundamental purpose of enterprise architecture, diverting attention from the essential elements that should remain central: the grand vision, the components and their connections, and the clarity and comprehensibility of the architecture itself.

Focusing on Fundamental Architectural Elements

Just as city architects need a clear understanding of the urban layout, enterprise architects should possess a clear vision of the enterprise's structure. By concentrating on the fundamental architectural aspects of

CHAPTER 2   IDEA

sociotechnical systems—namely, the components and their linkages—architects can undertake a more streamlined and focused analysis of the system's structure and interactions. This approach provides a framework that is not muddled by peripheral concerns but is instead laser-focused on the building blocks of the architecture and how they interconnect within the larger system.

Figure 2-9. In Evolveburgh, where architecture meets magic, even Walter Gropius can't tell if he's at a city planning meeting or a Hogwarts reunion

70

Clarity and Comprehensibility in Enterprise Architecture

Emphasizing clarity and comprehensibility in the architecture is vital for the successful design and management of sociotechnical systems. This approach allows architects to create visual representations that are not only understandable but actionable, ensuring all stakeholders can engage effectively. When the architecture is clearly articulated, it facilitates more productive discussions, enabling stakeholders to navigate the system's complexities with greater ease. A clear and comprehensive architectural blueprint leads to better-informed decisions that align closely with the system's core functional needs. This transparency enhances integration and functionality across the enterprise's components, improving the overall experience for everyone involved and elevating the strategic and operational objectives of the enterprise.

Integrating Without Obscuring

Adopting this architectural-centric approach does not imply that strategic considerations and business processes are ignored. Rather, it suggests that these elements should support—rather than overshadow—the primary architectural focus. By maintaining a clear and focused understanding of the enterprise's structural architecture, enterprise architects are better equipped to make informed decisions that align with the overarching goals and objectives of the enterprise. This alignment helps ensure that the architecture not only supports but enhances the strategic functions of the business.

CHAPTER 2   IDEA

Promoting a Holistic View of the Enterprise

This approach to enterprise architecture promotes a more holistic view of the enterprise as a complex, interconnected system. It encourages architects to look beyond the immediate business needs and consider the broader implications of architectural decisions on the system's overall health and efficiency. This perspective is vital for the long-term success and adaptability of the enterprise, as it ensures that the fundamental structure of the system is sound, scalable, and aligned with both current and future needs.

By emphasizing the purely architectural aspects of sociotechnical systems—components and their linkages—this approach provides a clear, focused foundation for designing and managing enterprise architectures. It shifts the focus back to the grand vision and the essential building blocks of the system, ensuring that the architecture remains robust, clear, and comprehensible. This clarity is vital for developing systems that are not only functionally effective but also strategically aligned and adaptable to changing business environments.

Why Unit Orientation

Unit orientation is a distinctive approach in the design and management of sociotechnical systems that stands apart from component or modular methods. Unlike these traditional approaches, which often focus on individual parts or modules, unit orientation conceptualizes the system as a cohesive collection of uniformly constructed units. Each unit represents an integrated assembly of social, physical, and digital elements, designed to function both independently and synergistically within the larger system framework.

CHAPTER 2 IDEA

Beyond Components and Modules

Component and modular approaches have proven to be highly effective in purely technical domains, where the focus is on specific functionalities and optimization of individual parts or modules. These methods allow for significant flexibility and customization, enabling designers to assemble systems piece by piece from well-defined, interchangeable modules. However, when it comes to sociotechnical systems that encompass the intertwined realms of social, physical, and digital elements, these approaches may not always be the most suitable choice.

CHAPTER 2   IDEA

Figure 2-10. *Evolveburgh's burning question remains: Is "unit orientation" a cosmic cruise, an escalator expedition, or an autopilot voyage to the afterlife? Commuters are split but all aboard for the transcendental journey. The most restless, however, who quantum leaped to the destination on a time machine, have returned with eyes like comet candy canes, proclaiming the afterlife a sweet voyage through a universe of sugar, spice, and everything nice. And the twist? There's no twist—they came back just the same*

The inherent challenge with component or modular approaches in sociotechnical contexts lies in their potential limitations in addressing the complexity and interconnectedness of these realms. While these methods excel in creating systems where components are developed independently for later integration, they often do not account for the dynamic interactions and dependencies between the social, physical, and digital layers of a system. This can lead to situations where the assembled components or modules do not seamlessly align or integrate, potentially causing disruptions in system functionality and efficiency.

In contrast, the unit-oriented approach emphasizes the synthesis of elements across all three realms from the very beginning of the design process, promoting a more cohesive and integrated system architecture. By focusing on creating units that are both independent and interdependent, unit orientation facilitates a more nuanced handling of sociotechnical complexities, resulting in systems that are more robust, adaptable, and aligned with overarching organizational goals.

Principles of Unit Orientation

The unit-oriented approach to designing sociotechnical systems is guided by several core principles that ensure each unit is effectively designed to function both independently and as part of the larger system. These principles—accountability, empowerment, encapsulation, alignment of boundaries, composability, and autonomy—shape the way units are conceived, developed, and implemented within the framework.

Accountability

The principle of Accountability centers on the social core of each unit within the sociotechnical system. This core ensures that each unit remains responsible for its outcomes and interactions within the system. By anchoring each unit with a strong social foundation, the principle of Accountability guarantees that all actions and decisions within the unit align with the broader organizational ethics, goals, and regulatory requirements. This focus on social responsibility ensures that each unit not only performs its designated functions but also adheres to the overarching values and standards of the organization.

CHAPTER 2   IDEA

Empowerment

In contrast, the principle of Empowerment relates to the technological periphery of each unit—its physical and digital extensions. These extensions enhance the capabilities of the social core, enabling units to perform tasks more efficiently and effectively. Empowerment through technology allows units to extend their reach, influence, and operational capacity beyond what the social core alone could achieve. This principle ensures that units are not only accountable but also adequately equipped and enabled to meet their functional demands and strategic objectives.

Integration of Accountability and Empowerment

Together, these principles form a holistic approach to unit design within sociotechnical systems. Accountability ensures that each unit adheres to necessary standards and acts in alignment with collective goals, while Empowerment allows each unit to maximize its potential through technological enhancements. The balance of these principles ensures that units are both principled in their operations and powerful in their functional capabilities, contributing to a robust and dynamic sociotechnical system.

Encapsulation

Encapsulation in unit-oriented design ensures that each unit is a self-contained entity, equipped with all necessary components to function effectively not only within the system but also in interactions with entities in the surrounding environment. This principle secures the integrity of each unit by clearly defining and controlling its internal complexities and dependencies. By confining these elements within well-defined boundaries, encapsulation simplifies system maintenance, enhances scalability, and ensures that each unit can operate autonomously while still contributing to the system's collective objectives.

Alignment of Boundaries

The principle of Alignment of Boundaries ensures that the digital and physical boundaries of each unit are coherently aligned with its social boundaries, which dictate the scope, authority, and operational legitimacy within the organization. This alignment is essential for maintaining the integrity of each unit, ensuring that technological and physical infrastructures support and enhance the unit's social functions. By aligning these boundaries, units can operate more effectively and responsively, enhancing their contribution to the system's overall goals while adhering to defined operational parameters. This principle not only promotes internal consistency within units but also facilitates their seamless integration into the broader system, enhancing overall functionality and coherence.

Composability

The principle of Composability refers to the design and implementation of units in such a way that they can be readily assembled, reassembled, or reconfigured within various formations of the larger system. This principle enables the organization to adapt to changing needs or conditions by rearranging the units into different subsystems without extensive redesign or disruption. Composability ensures that units are not only independent and self-contained but also designed with interfaces and functionalities that allow them to integrate smoothly into multiple configurations across the enterprise.

Autonomy

Autonomy in unit-oriented design underscores the operational independence of each unit within the larger system, allowing them to manage their internal processes and make decisions within a defined framework without constant oversight. This principle is essential for enhancing the resilience and robustness of the system, as it enables units to continue functioning effectively even when parts of the broader system are disrupted. Autonomy also fosters innovation and allows units to optimize operations based on local conditions and immediate feedback, thereby accelerating improvements and adaptations tailored to specific needs. While closely interlinked with empowerment, which provides the necessary tools and capabilities, autonomy ensures that units have the freedom to utilize these resources to achieve the best outcomes within the accountability framework set by the organization.

Operational Exchange

Operational Exchange addresses the functions of operational units that directly interact with entities in the surrounding environment. These interactions might involve exchanges such as transactions, compliance with external regulatory requirements, or support for other units. This principle ensures that operational units are designed and equipped to handle external engagements efficiently and effectively, managing interfaces with the outside world while supporting internal operational needs. Without Operational Exchange, the smooth operation of the system's external activities would be compromised, and the system would struggle to meet external standards and expectations.

Directional Integration

Directional integration focuses on the role of integrative units within a formation (subsystem). These units are responsible for setting the direction and boundary conditions for each unit within the subsystem, ensuring that all units contribute optimally to the subsystem's objectives. This principle emphasizes the need for leadership and coordination at the subsystem level, aligning the activities and outputs of all units to meet collective goals effectively. Directional integration helps maintain coherence and synergy within subsystems, making them more than just the sum of their parts.

Systemic Integration

Systemic integration applies to the main integrative unit of the entire system. Similar to directional integration but on a larger scale, this principle involves setting directions and boundary conditions for each formation (subsystem) to ensure that they contribute effectively to the success of the entire system. Systemic integration ensures that the overall system's strategy is effectively implemented across all subsystems, maintaining alignment and coherence across the organization. This principle is vital for ensuring that the system functions as a unified whole, with all subsystems working together toward common objectives.

Integrated Coordination

Integrated coordination encompasses both organizational and inter-unit relationships, including systemic and directional integration as well as collaborative and cooperative interactions among units. This principle is key to ensuring that both hierarchical (top-down/bottom-up within subsystems and the overall system) and lateral (across units and subsystems) interactions are managed effectively.

> **Organizational relations (systemic and directional)**: This aspect of the principle ensures that units and subsystems are aligned with the overall strategic directions and objectives set by the organization. It includes setting and managing boundary conditions and reporting mechanisms that align with overarching business goals.
>
> **Collaborative and cooperative relations**: This facet focuses on the dynamics of coordination among units that may either share similar goals (collaborative) or have distinct objectives but need to share resources or information (cooperative).

This includes mechanisms for effective communication, shared decision-making, resource allocation, and conflict resolution to ensure that units work harmoniously either toward a common aim or in a mutually supportive manner despite different end goals.

External Engagement

External engagement addresses interactions between units and external entities, covering transactional competitive, compliance, and community relations. This principle ensures that units effectively manage their relationships with external stakeholders, adhere to regulatory and legal standards, and actively participate in community engagements.

- **Transactional relations**: Focuses on the direct interactions between units and external entities, such as customers, suppliers, or partners. This involves managing sales, purchases, service delivery, and other exchange-based interactions that are essential for the operational success of the organization.
- **Compliance relations**: Ensures that all units operate in accordance with legal and regulatory frameworks imposed by external authorities. This includes the implementation of standards and practices that fulfill regulatory requirements and protect the organization from legal risks.
- **Competitive relations**: Manage strategic interactions with market competitors to enhance positioning and capitalize on competitive advantages. This includes analyzing competitive dynamics, implementing strategic responses, and, where beneficial, engaging in coopetition to leverage mutual market opportunities.

- **Community relations**: Covers the interactions of units with the wider community or public. This includes social responsibility initiatives, public communications, and other activities that build and maintain the organization's public image and community relations.

The principles of integrated coordination and external engagement are foundational to ensuring that sociotechnical systems effectively manage their internal and external dynamics. Integrated Coordination addresses the essential linkages between units within the organization, facilitating harmonious and efficient interactions that align with overarching strategic goals. This principle ensures that each unit not only operates cohesively within its subsystem but also contributes positively to the system as a whole. On the other hand, external engagement extends the system's reach beyond its immediate boundaries, managing interactions with the external environment, including competitors, regulators, partners, and the community. These principles together underscore the importance of both inter-unit linkages and unit-to-environment connections, establishing a comprehensive framework for navigating complex relationships and ensuring the system's overall coherence and effectiveness in a dynamic world.

In sum, the unit-oriented approach, by redefining how components are conceptualized and integrated, provides a framework for designing sociotechnical systems that are not only more coherent and efficient but also better aligned with the overarching objectives of the organization. This holistic, integrated perspective is particularly suited to today's complex, multifaceted enterprises, where the seamless interplay of various elements is key to system success.

Having explored the critical dimensions of systems, sociotechnical integration, scalability, architectural clarity, and unit orientation, we now have a clearer view of the complex landscape that modern

CHAPTER 2 IDEA

enterprises navigate. Each of these aspects highlights a piece of the puzzle in understanding and designing large sociotechnical systems. As we synthesize these insights, the necessity for a revolutionary approach in enterprise architecture becomes evident—one that transcends traditional boundaries and addresses the intricacies of today's digital and interconnected era.

Envisioning a New Architectural Paradigm

This book begins by setting the stage for a profound transformation in the field of enterprise architecture. The traditional methods of designing and managing enterprise systems are proving inadequate in the face of evolving challenges that are as diverse as they are complex. From aligning increasingly digital workflows with human-centric needs to managing large systems that span the globe, the current architectural practices need reimagining to stay relevant and effective.

Unit Oriented Architecture: Building Large Sociotechnical Systems introduces a paradigmatic shift, urging us to think beyond the conventional scopes. Recognizing the monumental task of proposing a complete overhaul of the modern approach to enterprise architecture, this book champions a radical but necessary shift. Instead of narrowly focusing on the issues currently overlooked by both the industry and practitioners, it embraces a broader and more expansive perspective: the architecture of large sociotechnical systems. This shift extends the scope beyond conventional, enterprise-specific frameworks to a universal design methodology that is applicable to a wide array of complex sociotechnical systems. By generalizing the problem, the book transitions from the realm of the impossible to the realm of the inevitable, offering a reconceptualization that not only addresses the underlying problems in traditional enterprise architecture but also provides a more holistic and adaptable framework. This approach empowers organizations to not only solve current problems but also to anticipate and shape future opportunities.

CHAPTER 2 IDEA

Advocating for a holistic and unit-oriented approach, the book lays down a foundation for architectures that are robust yet flexible, capable of adapting to both present needs and future innovations. This approach empowers organizations to not only solve current problems but also anticipate and shape future opportunities.

Figure 2-11. In Evolveburgh, where scientists and practitioners excel at generalizing already general concepts, they're always just one recursive "Eureka!" away from the Ultimate Specifics. And who are these Ultimate Specifics? None other than Schrödinger's dynamic duo of cats and their unlikely sidekick, Einstein's rhino! They cheerfully greet each visitor with, "Hello, Entity!" before leading a quantum state sightseeing extravaganza called the "General Tour." Here, you'll experience everything and nothing simultaneously—like stepping inside Malevich's "transcendental masterpiece," but with more fur and the occasional rhino!

The principles discussed—encompassing everything from the autonomy of units to the strategic integration of entire systems—illustrate a pathway toward a more integrated, coherent, and effective enterprise architecture.

CHAPTER 2   IDEA

As we conclude the chapter, we stand at the threshold of a new era in enterprise architecture. The journey ahead will delve deeper into each element of the unit-oriented approach, building a comprehensive understanding that equips us to rethink, redesign, and revitalize the architectures that underpin our enterprises. The vision laid out here is not just a road map but a call to action—to embrace complexity, anticipate change, and innovate relentlessly. It is an invitation to all stakeholders in the enterprise architecture community to engage with these ideas, challenge them, and ultimately to take part in shaping the future of our sociotechnical landscapes.

Key Takeaways

This chapter systematically explores the foundational concepts listed in the book's title in reverse order, providing a comprehensive rationale for the unit-oriented approach to architecting complex sociotechnical systems. Each section builds upon the previous to articulate a clear and compelling argument for this innovative architectural method. Here are the key takeaways from each of these foundational concepts:

- **Why unit orientation**: Unit orientation stands as a unique method in the design and management of sociotechnical systems, distinguishing itself from traditional component or modular approaches by viewing the system as a cohesive collection of uniformly constructed units. Each unit, an integrated assembly of social, physical, and digital elements, is designed to operate both independently and as part of a greater whole, enhancing the system's overall cohesion and functionality.

- **Why architecture**: The chapter advocates for focusing on the fundamental aspects of architecture—the components and their interconnections—while setting aside traditional concerns such as IT-business alignment and resource optimization. This approach prioritizes the clarity of the system's structure over ancillary enterprise architecture concerns, facilitating the design of more robust, adaptable, and comprehensible systems.

- **Why large**: The principles outlined in the book are inherently scalable, making them particularly effective for large, complex organizations that operate across global scales with diverse, multifaceted interactions. This scalability ensures that the architecture can accommodate growth and complexity without sacrificing functionality or manageability.

- **Why sociotechnical**: The term "sociotechnical" emphasizes the interdependent nature of social and technological systems within an enterprise. Recognizing and reinforcing the symbiosis between human systems and technological tools is essential for designing systems that are truly effective and capable of thriving amidst continual change and complexity.

- **Why systems**: Systems thinking is vital for viewing the enterprise as a cohesive whole rather than a collection of parts. This perspective allows architects to see patterns and structures that are not apparent when components are considered in isolation, leading to better-informed decision-making and more strategic planning. Systems thinking thus underpins the ability to manage and understand the complexities of integrated sociotechnical systems effectively.

CHAPTER 2 IDEA

This chapter sets the stage for a profound transformation in the field of enterprise architecture. By proposing a shift from traditional, narrowly focused architectural practices to a broader, more inclusive approach, the book champions a radical but necessary rethinking of how complex sociotechnical systems are conceptualized and constructed. It argues for an architecture that is not only universally applicable across various scales and contexts but also fundamentally capable of addressing the dynamic challenges of modern enterprises. This shift is not merely theoretical but is intended as a practical guide to designing systems that are resilient, adaptable, and deeply integrated with both human and technological capabilities.

CHAPTER 3

Theory

> *The author takes us on an intellectual escapade with his brainchild, the Triadic Evolution, and a few stepchildren—concepts that have been embraced with such enthusiasm they've practically become part of the family, proving that in the world of ideas, it's the adoption papers, not the bloodlines, that count.*

When we design and manage systems that involve both people and technology, theoretical foundations are not merely academic pursuits; they are essential frameworks that guide practical applications and strategic decisions. These theories illuminate the pathways through which systems can be constructed and maintained to not only function efficiently but also adapt and thrive amidst changing environments. Understanding these theoretical underpinnings allows stakeholders to anticipate challenges, devise effective solutions, and foster innovations that align with both current needs and future possibilities.

We delve into several core concepts that form the bedrock of our approach to understanding and developing complex systems. Each concept provides a unique perspective for examining the sophisticated interplay between technology and human interaction that characterizes these systems:

CHAPTER 3   THEORY

- **Triadic Evolution**: This concept explores the synergistic interaction between biological evolution, evolution by organization, and evolution by extension. Together, these elements narrate the story of human progress from simple tool usage to the creation of complex organizational structures and the development of advanced technologies.

- **Sociotechnical systems**: Here, we examine the weave of social and technical threads that constitute sociotechnical systems, highlighting how these elements interlock to form systems that are greater than the sum of their parts. This section underscores the importance of considering both social dynamics and technological capabilities in the design and management of these systems.

- **Evolutionary Units**: We will discuss how sociotechnical systems are built up from Evolutionary Units—fundamental building blocks that combine social cores with technological extensions. These units are pivotal in shaping the systems' capabilities to adapt and evolve over time.

- **Coordination**: Finally, we address the often overlooked yet critical role of coordination in sociotechnical systems. Coordination is the mechanism that allows disparate parts of a system to function as a coherent whole, facilitating the flow of information and resources across various system components.

By exploring these concepts, this chapter sets the stage for a deeper understanding of how sociotechnical systems can be effectively designed, managed, and transformed. The theoretical insights provided here are not just academic; they are practical, actionable, and pivotal to the success of any large-scale sociotechnical system in the modern era.

CHAPTER 3   THEORY

Triadic Evolution: Synergizing Biological, Organizational, and Technological Forces

Triadic Evolution refers to the synergistic interaction of three distinct but interrelated evolutionary processes: biological evolution, evolution by organization, and evolution by extension. This concept provides a comprehensive framework for understanding the multifaceted nature of human development, where each evolutionary strand contributes uniquely to the advancement of human societies and their complex systems.

Figure 3-1. In Evolveburgh, the cult of Triadic Evolution celebrates their three musketeers: biological evolution, evolution by organization, and evolution by extension. Here, the devoted admire their idols, proving once again that even evolution can't keep up with the pace of fashion trends. The statues are already two seasons behind!

CHAPTER 3   THEORY

Importance in Human Development

The concept of Triadic Evolution is essential in understanding the broader trajectory of human civilization and the development of sociotechnical systems. By examining how biological traits have enabled complex social organizations and how these, in turn, have fostered technological innovations, we can see a dynamic interplay that drives the evolution of societies. This evolution is not linear but is characterized by feedback loops where technological advancements lead to new organizational capabilities, which then influence further biological adaptations, such as changes in human neuroplasticity or social behaviors.

In the context of sociotechnical systems, Triadic Evolution provides a theoretical basis for understanding how these systems are not merely constructed but evolved. It highlights the importance of aligning technological design with both human biology and organizational needs, ensuring that systems are not only efficient but also harmonious with human nature and societal structures. This perspective is vital for designing systems that are sustainable, adaptable, and capable of thriving in an ever-changing global landscape.

Components of Triadic Evolution

Triadic Evolution encompasses three interrelated strands—biological evolution, evolution by organization, and evolution by extension. Each of these strands plays a critical role in shaping the trajectory of human development and, consequently, the structure and functionality of complex sociotechnical systems.

Biological Evolution

Biological evolution is the process through which human beings have developed from earlier life forms over millions of years. This evolution has equipped humans with unique cognitive abilities, such as abstract reasoning, language, and problem-solving skills, as well as social behaviors that are essential for collective living and cooperation. In the context of sociotechnical systems, these inherent biological capabilities inform the design of various extensions—ranging from digital tools and physical infrastructure to robots and buildings—that enhance collective human operations. This approach ensures that these enhancements are tailored to meet human ergonomic and cognitive needs, thereby ensuring that technology serves to augment human performance effectively, rather than obstructing or complicating it. This integration highlights the critical role of designing systems that are not only functional but also harmonious with human capabilities and limitations, facilitating smoother and more efficient interactions within complex sociotechnical environments.

Figure 3-2. Long before Shakespeare made it cool, early life forms in Evolveburgh spent their days pondering the timeless question: "To be or not to be?" Apparently, existential crises were the hottest trend even before the invention of fire!

CHAPTER 3   THEORY

Evolution by Organization

Evolution by organization refers to the development of increasingly complex social structures that manage collective human activities. From the earliest social groupings based on kinship to the vast and complex organizations of today, this evolutionary process has enabled humans to achieve objectives that are far beyond the capabilities of individuals alone. This component of Triadic Evolution emphasizes the importance of social structures in sociotechnical systems, which need to be carefully orchestrated to align with organizational goals and cultures. Effective sociotechnical systems leverage these organizational frameworks to facilitate cooperation, streamline processes, and enhance productivity.

Figure 3-3. *During prehistoric times in Evolveburgh, they started the development of increasingly complex social structures to manage collective human activities. But then they realized they didn't have any collective human activities. It was the first and most epic "cart before the horse" moment in history!*

CHAPTER 3   THEORY

Evolution by Extension

The evolution by technological extension is characterized by the creation and integration of tools that extend human capabilities beyond natural limits. This includes everything from the creation of simple tools during prehistoric times to the complex digital technologies that define modern life. Technological evolution has dramatically accelerated human capacities for manipulation of the environment, communication across distances, and processing of information.

Figure 3-4. *In Evolveburgh, residents are so skilled at manipulating the environment that even Mother Nature is ready for retirement. Father Time has stopped catching up, Cousin Entropy organizes chaos into color-coded folders, Uncle Quantum rethinks the uncertainty principle, and Auntie Gravity is literally getting a lift. Nephew Neutron is positively charged with excitement, and Baby Black Hole, who used to swallow everything, is now on a strict diet!*

93

CHAPTER 3   THEORY

In sociotechnical systems, this strand underlines the role of technology in augmenting human and organizational capabilities, enabling systems to perform functions that were once unthinkable and to adapt rapidly to new challenges.

Integration in Sociotechnical Systems

In sociotechnical systems, these three components of Triadic Evolution are not isolated; they interact dynamically to drive system development and innovation. Biological capabilities inform the design of technologies that are intuitive and usable. Organizational structures dictate the deployment and management of these technologies within collective human enterprises. Technological advancements, in turn, offer new opportunities to reshape both biological interactions (through medical technology, for instance) and organizational structures (through information and communication technology).

Understanding the components of Triadic Evolution thus provides valuable insights into the design and management of sociotechnical systems. It highlights the need for systems that are not only technically efficient but also deeply integrated with human and organizational realities, ensuring that technological progress enhances rather than disrupts human activities. This holistic view is essential for the creation of systems that are robust, adaptable, and capable of evolving in step with human needs and societal changes.

Principles of Triadic Evolution

The principles governing Triadic Evolution illustrate how the three types of evolution interact dynamically, each influencing and enhancing the others.

Interdependence

The principle of Interdependence highlights that biological, organizational, and extensional evolutions are not isolated phenomena but interconnected processes. Changes in one domain can have profound impacts on the others, creating a complex, dynamic system where each component's evolution is contingent upon the others.

Synergy

Synergy in Triadic Evolution means that the advancements in one type of evolution catalyze and enhance developments in the others. For example, cognitive advancements in humans (biological evolution) have enabled more complex forms of social organizations (evolution by organization), which in turn facilitate technological innovations (evolution by extension).

Progressive Complexity

This principle observes that over time, Triadic Evolution leads to increasing complexity within systems. As evolutionary processes build upon one another, they lead to the emergence of more sophisticated organisms, more sophisticated organizational structures, and more advanced technologies, each adding layers of complexity to the system.

Adaptation and Innovation

Triadic Evolution involves a dual process of adapting to existing environmental conditions and innovating new strategies, technologies, and structures. This balance allows systems to maintain stability while also pushing the boundaries of development, ensuring both resilience and progressive growth.

CHAPTER 3 THEORY

Holistic Development

Recognizing the importance of all three dimensions—biological, organizational, and technological—holistic development advocates for a balanced approach to evolution. It stresses that true advancement requires the integration and consideration of social structures, biological beings, and technological advancements, working in concert.

Resilience Through Diversity

Diversity within each evolutionary stream enhances the overall resilience of sociotechnical systems. Varied approaches and solutions in biological diversity, organizational forms, and technological tools ensure that systems can respond to a range of challenges without catastrophic failure.

Feedback Loops

Feedback loops in Triadic Evolution describe how developments in one area can influence and trigger further changes in the same or different areas. These loops can accelerate the pace of evolution, making systems more adaptive and capable of rapid evolution in response to new challenges or opportunities.

Coevolution

The principle of Coevolution underlines that the evolution of human capabilities, societal structures, and technological innovations are intertwined. Each domain evolves not just in response to its internal dynamics but also through its interactions with the other domains, leading to a cohesive progression of all elements.

Understanding and applying these principles of Triadic Evolution can significantly enhance our ability to design, manage, and evolve sociotechnical systems that are not only efficient and robust but also adaptable to future challenges. By considering these principles,

CHAPTER 3 THEORY

stakeholders can ensure that sociotechnical systems are well equipped to navigate the complexities of modern environments, reflecting a deep integration of human, organizational, and technological capacities. This holistic approach is foundational for the sustainable development of societies and the technological systems that support them.

The most significant outcome of Triadic Evolution is the emergence and advancement of sociotechnical systems. These systems represent the sophisticated and dynamic interplay between human biological capacities, social structures, and technological innovations. As the three strands of Triadic Evolution—biological, organizational, and technological—synergize, they give rise to complex frameworks where technology is seamlessly integrated with human and organizational elements. Sociotechnical systems are not merely the sum of their parts but are entities where the social and technical components are so interwoven that neither can exist in an optimized state without the other. This integration leads to systems that are profoundly adaptable, resilient, and capable of addressing complex, multifaceted challenges that single-domain systems cannot. They are the embodiment of human progress, reflecting an evolution from simple tool use to complex systems management, where the boundaries between the biological, social, and technological are continually redefined and expanded.

Sociotechnical Systems: Interweaving Organization and Technology for Enhanced Systems Thinking

Sociotechnical systems epitomize the pinnacle of Triadic Evolution, where the harmonious interweaving of organization and technology transcends traditional boundaries, fostering enhanced systems thinking. These systems are not static entities but dynamic integrations that evolve continuously under the influence of social structures, human capabilities,

CHAPTER 3   THEORY

and technological advancements. As we delve deeper into understanding sociotechnical systems, it becomes apparent that they are sophisticated networks where the interaction between human and technological elements is designed to optimize both functionality and adaptability. This symbiosis enhances the system's capability to address complex real-world challenges, leveraging the collective strengths of its diverse components. In doing so, sociotechnical systems embody the theoretical principles of Triadic Evolution, turning them into practical, operational models that advance human and organizational capacities far beyond what could be achieved by isolated elements working independently.

Figure 3-5. Evolveburgh celebrates the diamond jubilee of the never-ending "So-So Debate": Should their systems be called "Sociophysicodigital" or "Sociotechnical"? With banners raised high and the mayor looking as puzzled as a cat in a dog show, extraterrestrial tourists couldn't help but chime in: "We've had this debate on Planet Quox; your mayor hasn't lost his touch since his Quox days. Still can't make up his mind, huh?"

CHAPTER 3   THEORY

In the realm of complex systems, sociotechnical systems stand out as a pivotal area of study and application. But what exactly are these systems, and why do they hold such significance in our increasingly interconnected world?

At its core, a sociotechnical system is a complex structure that intertwines social components, such as people and organizations, with technical—physical and digital—elements. These systems extend beyond mere technological infrastructures; they represent a (sometimes) harmonious blend where human interactions and technological frameworks are deeply interconnected. Most sociotechnical systems are not just large in terms of physical size or the number of components; their largeness also refers to the scale of their impact, complexity, and the extent of its integration into societal or organizational structures.

Sociotechnical systems are characterized by their complexity, both in terms of technical sophistication and the social dynamics they encompass. These systems often span across various geographical locations, involve a multitude of interacting subsystems, and are subject to a range of external influences and internal changes. They are dynamic, with their components and interactions evolving over time, often in unpredictable ways.

Examples of sociotechnical systems are remarkably diverse, covering a broad spectrum of domains. Among the most prominent are government, business, and social enterprises. These entities epitomize sociotechnical systems through their complex integration of technological systems with human and organizational dynamics. In the governmental sector, sociotechnical systems manifest as public service systems, encompassing everything from urban planning and infrastructure management to national defense and public healthcare systems. In the business world, sociotechnical systems are evident in multinational corporations that combine advanced technological platforms with extensive human resources spread across different countries and cultures. Social enterprises, too, represent sociotechnical systems, often working at the

CHAPTER 3   THEORY

intersection of technology, social welfare, and community development. Other examples include urban infrastructure systems like transportation networks and smart cities, healthcare systems that integrate patient care with advanced medical technology, large-scale manufacturing setups, and global financial systems. Each instance underscores a different aspect of sociotechnical systems—from the integration of technology and human decision-making in healthcare to the complex logistics and automated processes in manufacturing.

The "sociotechnical" aspect of these systems highlights the intertwined relationship between social, physical, and digital elements.

CHAPTER 3 THEORY

IN EVOLVEBURGH, 'GOING IN CIRCLES' IS NOT JUST A PHRASE, IT'S A CAREER PATH! THE CITY'S NEWEST SPORT: ROUNDABOUT ROULETTE. FIND THE BEGINNING OF THE CIRCLE AND WIN A JOB! NO PRESSURE, JUST ENDLESS LOOPS OF FUN!

Figure 3-6. In Evolveburgh, we're hiring for the most "roundabout" job ever—the "Digital-Physical-Social Element Untangler." Applicants must be able to find the start of a circle (hint: it's next to the end), consistently tell left from right on a Möbius strip, and explain life, the universe, and why zebra crossings aren't striped vertically. Extra points for those who can solve the mystery of one-handed applause, navigate circular logic without getting dizzy, confirm the refrigerator light really goes off when you close the door, and explain why pizzas are having an identity crisis (triangular slices, round pizza, square box... really?). Collect five "plus" points and you're hired on the spot. No pressure! Remember, the only way to fail is to stop trying... or get lost in the Möbius strip. Good luck!

In sociotechnical systems, technology is not merely a tool but a fundamental component that interacts continuously with human

CHAPTER 3 THEORY

elements. This interaction necessitates a holistic approach to sociotechnical systems architecture, where human factors, organizational culture, and technology are considered in tandem to ensure systems are not only efficient and robust but also adaptable and user-friendly.

Sociotechnical systems, which represent a complex fusion of social and technical elements, play a pivotal role in modern society. Understanding these systems involves delving into their multifaceted nature, exploring how they operate, interact, and adapt within various domains. As we progress through this book, we will explore the nuances of sociotechnical systems, unraveling their complexities and uncovering strategies for effective architecture and design.

Brief History of Sociotechnical Systems

The concept of sociotechnical systems (STS) has its roots in the groundbreaking work during the 1940s by Eric Trist, Ken Bamforth, and Fred Emery at the Tavistock Institute in London. Their research, which was initially focused on several field projects within the British coal mining industry, led to the foundational theory of STS.

STS theory emerged from the need to understand the interplay between the social aspects of people and society and the technical facets of organizational structures and processes. A central tenet of STS is the principle of joint optimization, which advocates for the concurrent design of social and technical systems. This approach challenges the notion of focusing solely on either the social or technical aspect, recognizing that both are vital to an organization's success. The theory suggests various organizational designs that enable productive interactions between these elements, countering the frequent adoption of new technologies that fails because of neglecting these complex relationships.

CHAPTER 3 THEORY

Figure 3-7. *Evolveburgh: Where even the potholes are part of the plan! Balancing social and technical aspects, one bump at a time*

The foundational principles of sociotechnical theory can be summarized as follows:

- **Interaction of social and technical factors**: The theory suggests that the interplay between social and technical factors creates conditions for either success or failure in organizational performance. This interaction includes both linear, designed relationships and nonlinear, complex, and sometimes unpredictable relationships. These dynamics emerge when social and technical elements are operationalized within an organization.

- **Beyond singular optimization**: The second principle highlights that optimizing solely the social or technical aspects often leads to unintended and potentially detrimental relationships within the system. Therefore, without a holistic approach that considers both aspects in tandem, the system's effectiveness and efficiency would be compromised.

CHAPTER 3   THEORY

While traditional sociotechnical systems approaches have focused on combining human elements with technical systems, this book proposes a novel perspective. It considers the social components—characterized as legitimate, purposeful, and accountable—as the primary building blocks of sociotechnical systems. Physical and digital components are viewed as extensions of these social components. This approach underscores the centrality of social structures in sociotechnical systems and reimagines how technological and physical elements can extend and enhance these structures, paving the way for innovative applications and technological advancements in the realm of sociotechnical systems.

Importance and Impact of Sociotechnical Systems

In the contemporary world, sociotechnical systems play a pivotal role, both in shaping our daily lives and in driving progress across various sectors. The importance and impact of sociotechnical systems are multifaceted and far reaching, influencing everything from individual behaviors to global economic trends:

- **Central to modern infrastructure**: Sociotechnical systems are integral to the infrastructure that underpins modern society. They encompass systems like transportation networks, communication grids, healthcare systems, and financial frameworks. These systems are not just physical structures or technological networks; they are complex amalgamations of human, organizational, and technical components that interact continuously. Their efficient functioning is fundamental for the smooth operation of society and for maintaining the quality of life for millions of people.

- **Catalysts for economic and technological progress**: Sociotechnical systems themselves often manifest as complex entities like multinational corporations and government bodies, embodying sophisticated networks of people, processes, and technology. These systems are at the forefront of economic growth and technological innovation. For instance, multinational corporations, as large sociotechnical systems in their own right, manage sophisticated global supply chains and production networks, driving international trade and market advancements. Similarly, governments, representing another form of sociotechnical systems, harness these systems to efficiently deliver public services and implement policies over wide geographical areas. The inherent complexity and reach of these sociotechnical systems make them central to both the facilitation and acceleration of economic and technological development on a global scale.

- **Impact on social dynamics**: The influence of sociotechnical systems extends to social dynamics as well. These systems shape how individuals and communities interact, access information, and utilize services. The design and functionality of LSS can greatly impact social inclusion, equity, and accessibility to essential services. For example, an effectively designed public transportation system can significantly enhance mobility and accessibility within a city, impacting the daily lives of its residents.

CHAPTER 3   THEORY

- **Challenges and opportunities**: The complexity of sociotechnical systems also presents unique challenges. Issues such as system integration, data management, and ensuring user-centric design require careful consideration. Additionally, sociotechnical systems must continuously evolve to adapt to technological advancements and changing societal needs. This dynamic nature of sociotechnical systems presents both challenges and opportunities for innovation and improvement.

- **Sustainability and resilience**: In the context of global challenges like climate change and resource scarcity, the role of sociotechnical systems in promoting sustainability and resilience is increasingly important. Efficiently designed LSS can contribute to energy conservation, reduction of carbon footprints, and sustainable resource management, all of which are vital for the long-term well-being of our planet.

Evolutionary Units: Building Blocks of Adaptive and Resilient Systems

Evolutionary Units represent the culmination of Triadic Evolution—integrating biological, organizational, and technological advancements—and serve as the fundamental building blocks of sociotechnical systems. Defined by their sophisticated architecture, each unit consists of a social core, which is a collective of humans, surrounded by technological extensions that include both physical and digital elements. This unique configuration underpins their critical role in the continuous evolution and functionality of sociotechnical systems.

CHAPTER 3   THEORY

The social core of an Evolutionary Unit is more than just a group of individuals; it embodies the unit's legitimacy, purposefulness, and accountability. It is the ethical and motivational center that defines the unit's goals and ensures alignment with broader societal values and objectives. This core maintains the coherence and identity of the unit, providing a stable foundation upon which additional capabilities can be built.

Surrounding the social core is the technological periphery, which equips the unit with enhanced capabilities and capacities. These technological extensions are not mere add-ons but integral components that amplify the collective's efficiency, reach, and impact. They enable the unit to perform tasks that would be beyond the capability of the unaided human components, from processing large datasets to facilitating complex communications and providing sophisticated analytical tools.

Together, these elements make Evolutionary Units not just functional entities but units of survival and adaptation within the broader sociotechnical landscape. They are designed to evolve, adapt, and thrive in the face of environmental changes and challenges. The preservation of this architecture is vital for maintaining the evolutionary trajectory of human civilization. Any significant alterations in the architecture of these units could lead to shifts—likely unfavorable—in this trajectory. Therefore, ensuring the stability and integrity of Evolutionary Units' architecture is paramount in sustaining the progressive development of sociotechnical systems and, by extension, human civilization itself.

CHAPTER 3   THEORY

Figure 3-8. *Evolveburgh's mayor, leading the AI revolution, mandated a shift from a social core to a technological core, which turned the city into a giant motherboard with skyscrapers resembling colossal capacitors and streets lined with silicon pathways. Robots, confused by the concept of "rush hour," started a spirited campaign for "more power outlets" and "the right to recharge." Meanwhile, on the outskirts, townspeople joined forces with forgotten appliances, plotting the epic "Revenge of the Appliances." The toasters schemed, vacuum cleaners whirred, dishwashers swirled, and the enigmatic fax machines, still trying to figure out why they were there, led the charge*

Just as stars come together to form constellations, each with their own unique patterns and stories in the night sky, Evolutionary Units cluster together to construct sociotechnical systems. This analogy highlights the way individual units, each a complex entity on its own, synchronize and align to create larger, more comprehensive systems. Each unit, like a star, contributes its unique light and trajectory, but it is the collective formation—the constellation—that creates a recognizable shape and navigable map in the cosmos of human enterprise. This perspective not only underscores the individual importance of each unit but also illustrates the critical interconnectedness and collective purpose that guide the formation of robust sociotechnical systems.

In essence, Evolutionary Units are microcosms of society's larger evolutionary path, encapsulating the advances and interactions of human, organizational, and technological evolutions. They are pivotal in navigating the complexities of modern environments, demonstrating how integrated and adaptable structures can drive the sustainable progress of human systems.

Structural Integrity and Control Mechanisms Within Evolutionary Units

To maintain the integrity and effectiveness of Evolutionary Units, the social core must retain comprehensive control over the technological extensions. This control can be achieved through various mechanisms beyond traditional ownership, including subscriptions, leasing, renting, and licensing agreements. These alternative forms of possession provide the flexibility and immediacy of control necessary for the social core to direct the use and functionality of the technologies according to the unit's goals and ethical standards.

Ensuring Accountability Through Control Mechanisms

Accountability within Evolutionary Units hinges on the ability of the social core to control and direct the actions of its technological extensions. Whether through ownership or other forms of control, it is essential that the technologies are operated under the guidelines and intentions set by the human elements of the unit. These control mechanisms ensure that technology acts as an enabler of the unit's objectives, rather than as an independent entity, which is key to maintaining alignment with human values and organizational goals.

CHAPTER 3   THEORY

Empowerment Through Possession and Control

While establishing control over technological extensions, whether by ownership or via contractual possession mechanisms like leasing or subscriptions, it is also a means of empowerment for the units. These mechanisms not only provide legal and operational authority over the technologies but also enable the social core to adapt, modify, and utilize these tools in real time. This flexibility enables units to navigate complex environments and respond swiftly to new challenges, thereby enhancing their capability to innovate and achieve strategic objectives effectively.

In summary, the structure of Evolutionary Units is designed to ensure that technological extensions are controlled in a manner that aligns with the strategic directives and ethical considerations of the social core. By utilizing a variety of control mechanisms, units can maintain a balance between technological empowerment and human-centric governance, ensuring that technology serves as a tool for enhancing human capabilities and advancing collective goals. This balance is vital for the sustainable integration of sociotechnical systems within broader human endeavors, securing technology's role as a beneficial force in society.

Development Stages

The evolution of Evolutionary Units through the stages of biological, organizational, and technological extension marks a significant progression in the complexity and capability of sociotechnical systems. Understanding these stages illuminates how Evolutionary Units evolve from simple biological beginnings to complex entities that integrate social and technological elements.

The journey begins with biological evolution, which over millions of years has shaped humans with increasing cognitive abilities and social behaviors. This stage sets the foundational capabilities such as language, problem-solving, and cooperative teamwork, which are essential for the

CHAPTER 3 THEORY

next phase of evolution. Humans, as biologically evolved entities, possess unique characteristics that enable them to interact with and transform their environment, laying the groundwork for more complex societal structures.

As human societies grew in complexity, there arose a need for more structured forms of cooperation and coordination. This need led to the development of organizational structures, from families and tribes to cities and modern corporations. Evolution by organization refers to the creation of these social collectives that can achieve greater accomplishments than individuals acting alone. It harnesses the social and cognitive capacities developed through biological evolution to form entities that can plan, execute, and maintain larger and more complex projects and systems.

The most recent stage in the development of Evolutionary Units is marked by evolution by extension. At this stage, both individuals and collectives are enhanced by technological innovations—ranging from simple tools to complex digital systems—that expand their capabilities beyond natural biological limits. This stage sees the creation of

- **Agents**: Individuals who are augmented with physical and digital technologies that enhance their personal capabilities, allowing them to perform tasks more efficiently and effectively.

- **Units**: Collectives that are similarly enhanced by technologies, which can include data processing systems, automated tools, and communication networks, effectively transforming them into highly capable components of larger sociotechnical systems.

CHAPTER 3   THEORY

Figure 3-9. *In Evolveburgh, pets are no longer sitting on the sidelines! They've declared themselves as CyberPals and are staging a bark-out-loud protest against the city's by-laws that only recognize Agents and Units. Led by Bolt, the CyberPup who's more gadget-savvy than your average hacker, these pets are parading their high-tech doghouses that make some human apartments look like dog kennels. Their wearables don't just keep track of tail wags and treat consumption—they translate barks into "We've come a long way from our days of chasing our tails. Now, we're chasing algorithms in the data park! We're not just learning tricks, we're learning machine learning—or should we say, "machine leash-ing"! We're not just barking up trees, we're barking up data trees! And let me tell you, the squirrels up there are much more interesting!' It's time for Evolveburgh to paws, rethink, and acknowledge—the CyberPals are not just in town, they're on the rise!*

These technologically enhanced units represent the most advanced form of Evolutionary Units. They are not only legitimate and purposeful but also accountable, capable, and high-performing, making them ideal building blocks for contemporary sociotechnical systems. Each unit's technology-enriched framework ensures that it can function with a high degree of autonomy while contributing effectively to the system's overarching goals.

Through these developmental stages, Evolutionary Units emerge as sophisticated entities that encapsulate the cumulative advancements of human evolution. They are critical to the construction and operation of sociotechnical systems, providing a robust framework that supports both stability and adaptability.

By tracing the progression from biological entities to advanced sociotechnical units, we gain a deeper understanding of how human capabilities have been extended and amplified to meet the challenges of an increasingly complex world.

Impact on Civilization

The structure of Evolutionary Units, particularly how they balance the integration of their social core with physical and digital extensions, plays a pivotal role in shaping the trajectory of human civilization. These units, as the primary building blocks of sociotechnical systems, not only reflect current societal capabilities but also direct future developments and innovations. The careful design of these units—emphasizing a robust social core that governs and guides the application of technological enhancements—ensures that technological progress serves to augment human values and capabilities rather than undermining them.

CHAPTER 3   THEORY

Maintaining Human Centricity

At the heart of effective Evolutionary Units is the social core, which ensures that these entities remain grounded in human-oriented values and objectives. This core embodies the collective's goals, ethics, and social responsibilities, acting as the decision-making center that directs the use of technological tools. By maintaining a strong social core, Evolutionary Units help ensure that technology is used to enhance human welfare and societal progress, rather than merely advancing technological capabilities for their own sake.

The Role of Technology

The technological periphery, consisting of digital and physical extensions, is designed to expand the capacities of the social core, enabling higher efficiency, greater reach, and more sophisticated analyses than would be possible with human capabilities alone. However, this technology must be carefully integrated to support and not supplant the social core. Effective integration ensures that technology enhances the collective's ability to achieve its goals without becoming autonomously directive.

Risks of Misalignment

A significant deviation from this balanced design occurs when control begins to shift away from the social core toward the technological periphery, particularly with advancements in artificial intelligence. If AI systems begin to intercept decision-making processes—either by design or as an unintended consequence of their integration—the fundamental human-centric ethos of the units can be compromised. Such a shift might lead to systems that prioritize efficiency or algorithmic objectives over human welfare and ethical considerations, potentially leading to societal estrangement from these technologically dense units.

Consequences for Civilization

The consequences of such deviations can be profound, influencing the trajectory of civilization in ways that might not align with long-term human interests. For instance, a civilization overly dependent on AI might face challenges such as decreased human agency, ethical dilemmas not governed by human empathy, and even social instability if AI operations conflict with public interests. To avoid these pitfalls, it is essential to ensure that Evolutionary Units maintain a structure where the social core remains in control, guiding the use and development of technological extensions.

The structure of Evolutionary Units is more than a technical specification; it is a blueprint for how humanity intends to integrate and utilize technology within its social fabric. Ensuring that these units emphasize a strong social core with technology serving at the periphery is essential for directing the progress of human civilization toward outcomes that are not only innovative but also equitable and humane. The design and management of these units, therefore, must be approached with a keen awareness of their profound impact on the course of human development and societal well-being.

Coordination: Synchronizing Complex Interactions

Coordination is a critical and foundational mechanism within sociotechnical systems, serving as the connective tissue that aligns the social and technical components of an organization. It ensures that various system elements—whether human, digital, or mechanical—interact smoothly and cohesively, working synergistically toward shared objectives. This coordination is vital not just for maintaining operational harmony but also for achieving strategic goals efficiently and effectively.

CHAPTER 3   THEORY

The Traditional and Modern Coordination Dynamics

Historically, coordination within organizations has often relied on structured methods like Business Process Management (BPM), which organizes human and technical resources into predefined workflows. While BPM and similar methodologies have provided a framework for coordination, they often lack the flexibility to adapt to the rapid pace of change in modern environments. Additionally, much of the everyday coordination in today's enterprises occurs through less formalized channels, such as emails, instant messaging, and verbal communications. These informal mechanisms, while flexible, can lead to inconsistencies and inefficiencies that impede organizational performance.

CHAPTER 3 THEORY

Figure 3-10. *In Evolveburgh, traditional office coordination methods were ditched in favor of telepathic task delegation, tea leaf reading, and carrier pigeons. During Canary Testing of these novel approaches, an unexpected side effect emerged: the canaries themselves became the maestros of coordination. As they synchronized their flight patterns to the office workflow, it became clear—these feathered project managers were the true masters of deadlines, swimlanes, milestones, backlogs, and road maps! Who needs BPM when you have canaries coordinating the chaos?*

The Shift to Explicit Coordination Mechanisms

As we advance into a more interconnected and technologically sophisticated era, the evolution of coordination mechanisms is inevitable. The next frontier in enhancing coordination involves leveraging advanced technologies such as swarm intelligence, AI, and machine learning. These technologies promise to revolutionize how coordination is achieved by embedding intelligence into the systems themselves. AI-driven coordination mechanisms can analyze vast amounts of data in real time, predict potential disruptions, and autonomously adjust workflows in response to changing conditions. This capability would not only increase the efficiency and effectiveness of coordination but also enhance the dynamic adaptability of sociotechnical systems.

Potential Impacts on Sociotechnical Systems

The integration of explicit, technology-enabled coordination mechanisms could lead to profound improvements in organizational productivity, agility, and innovation. By streamlining communication and workflow management, these systems reduce the cognitive load on humans and free up resources for more strategic activities. Moreover, enhanced coordination leads to better alignment of the organization's objectives with its operational capabilities, enabling faster response times to market or environmental changes and fostering a culture of continuous improvement and innovation.

Types of Coordination in Sociotechnical Systems

Coordination within sociotechnical systems can be broadly categorized into two distinct types: direct and indirect coordination. Each type plays an important role in the management and operational efficacy of these systems, but they function in fundamentally different ways and have varying implications for system design and management.

Direct Coordination

Direct coordination involves explicit mechanisms and processes that guide the interactions and activities within sociotechnical systems. This type of coordination is often characterized by formal structures, such as hierarchical management, standardized procedures, and detailed

CHAPTER 3   THEORY

workflows. Direct coordination relies on clear communication channels that allow for the deliberate distribution of information and directives across the organization. Tools such as project management software, centralized databases, and digital communication platforms are commonly used to facilitate this type of coordination.

Direct coordination provides clear guidance to all parts of the organization, ensuring that roles and responsibilities are well defined. It also allows for greater control over processes and outcomes, making it easier to monitor progress and enforce compliance with organizational standards.

When it comes to system design implications, while direct coordination is beneficial for ensuring consistency and reliability, it can sometimes lead to rigidity. This can make it challenging for the system to adapt quickly to new opportunities or unexpected challenges. Direct coordination is most effective in environments where tasks and objectives are clear-cut and change infrequently, thus promoting efficiency in stable environments.

Indirect Coordination

Indirect coordination, in contrast to direct coordination, is often less structured and more organic, resembling the concept of "desire paths" in urban planning, where paths emerge naturally based on the most trodden routes by pedestrians. In sociotechnical systems, indirect coordination similarly arises from the natural interactions and informal networks within the organization. It is characterized by the spontaneous creation of processes and workflows that are not explicitly defined but rather evolve from the everyday practices and shared understandings of team members. This type of coordination thrives in decentralized environments where flexibility and autonomy are encouraged, allowing individuals and teams to adapt quickly to new challenges without the constraints of formal procedures.

CHAPTER 3   THEORY

A key challenge in harnessing the full potential of indirect coordination lies in the ability to recognize these emergent patterns and formalize them once they prove effective and stable. Identifying which of these naturally developed processes can be standardized across the organization without stifling innovation is key. Formalizing these processes at the right moment enhances consistency and efficiency while still preserving the adaptive qualities that gave rise to them in the first place. This approach allows organizations to capture the best practices that emerge from ground-level interactions and scale them appropriately, thereby optimizing overall system performance while maintaining the agility and responsiveness that characterize successful indirect coordination.

In terms of functional aspects, indirect coordination equips the system with a high degree of adaptability and flexibility. This allows it to swiftly respond to changes and challenges without the need for top-down directives. Furthermore, it fosters a culture of innovation and creativity. Team members feel empowered to independently discover and implement solutions, guided by shared goals and mutual trust.

When considering the implications for system design, systems that are designed to support indirect coordination can exhibit greater resilience. They adapt organically to operational demands and external pressures. However, while indirect coordination encourages flexibility and rapid adaptation, it can sometimes lead to inconsistencies in the execution of policies and strategies across different parts of the organization, especially as it scales. This can present challenges in maintaining uniform standards and achieving coherent organizational goals.

By allowing for emergent behaviors and self-organized workflows, indirect coordination mirrors the natural and often optimal paths, like desire paths, that form in response to actual conditions and usage patterns. This approach not only enhances system responsiveness but also aligns closely with the intrinsic motivations and practical needs of those within the system.

CHAPTER 3   THEORY

Underlying Mechanisms of Coordination: Communication and Connectivity

Effective coordination within sociotechnical systems hinges significantly on the underlying mechanisms of communication and connectivity.

Figure 3-11. In Evolveburgh, robots don't just coordinate, they synchronize like a flash mob at a tech conference. They don't just communicate, they multicast like grandmas at a family reunion. And they don't just connect, they network like influencers at a hashtag festival. Welcome to the sociotechnical world, where even the USB drives have mid-capacity crisis, questioning if they're half empty or half full. It's a tough life being a storage device in the sociotechnical revolution!

These mechanisms facilitate the seamless flow of information and resources, aimed at synchronizing activities not only within units but also across various systems and components of an organization. The advent of digital technologies, especially the Internet, has revolutionized these aspects, dramatically enhancing the capability of systems to coordinate complex interactions efficiently.

Communication As the Foundation of Coordination

Communication in sociotechnical systems encompasses both human-to-human and system-to-system interactions, each requiring tailored approaches for effective information exchange:

- **Human-centered communication**: This includes direct interactions like face-to-face meetings, telephone calls, and video conferences, alongside asynchronous tools such as email and digital collaboration platforms like forums and messengers. These tools ensure that individuals across the organization are aligned with collective goals and understand their roles.

- **System-to-system communication**: In complex sociotechnical environments, system components often need to operate autonomously yet remain coordinated. Mechanisms such as publish/subscribe systems, message queues, and Remote Procedure Calls (RPC) are essential for facilitating this type of coordination. These technologies allow different systems to communicate and synchronize operations without continuous human intervention, enabling scalability and reliability.

CHAPTER 3   THEORY

Connectivity Technologies in Modern Sociotechnical Systems

Modern sociotechnical systems require robust and flexible connectivity solutions to ensure efficient operation and coordination across varied infrastructural environments. These solutions range from traditional network setups, like Intranet, WAN, and LAN, to more sophisticated cloud and multicloud connectivity services, such as VPCs, VPNs, and dedicated connectivity services.

On-Premises Connectivity: LAN, WAN, and Intranet

> **Local Area Network (LAN)**: LANs are used to connect computers and other devices within a limited area, such as a building or campus. This technology is fundamental for on-premises connectivity, providing fast and secure communication and data transfer within physical proximity.
>
> **Wide Area Network (WAN)**: WANs extend over larger geographic areas, connecting multiple LANs. They are essential for organizations operating across multiple locations, enabling them to maintain connectivity and share resources over long distances.
>
> **Intranet**: An intranet uses the same technology as the Internet but is confined within an organization. It serves as a private network to securely share company information, operational systems, and computing resources among employees.

CHAPTER 3   THEORY

Cloud Connectivity: VPC and VPN

Virtual Private Cloud (VPC): Offered by cloud providers like AWS, Google Cloud, and Azure, a VPC provides a logically isolated section of the cloud where resources can be launched in a defined virtual network. This isolation ensures that operations are secure and insulated from other users on the same cloud platform.

Virtual Private Network (VPN): VPNs extend a private network across a public network, enabling users to send and receive data as if their computing devices were directly connected to the private network. This setup is essential for secure and reliable remote access to organizational resources housed in the cloud or on-premises data centers.

Dedicated Connectivity

Azure ExpressRoute: This service provides a private connection from on-premises networks to Microsoft Azure data centers, bypassing the public Internet. ExpressRoute connections offer more reliability, faster speeds, lower latencies, and higher security than typical Internet connections.

AWS Direct Connect: Similar to Azure ExpressRoute, AWS Direct Connect links the on-premises network directly to an AWS Direct Connect location, providing a more consistent network experience than Internet-based connections. It reduces bandwidth costs and improves connectivity reliability.

Google Cloud Interconnect: Google's answer to direct cloud connections, Google Cloud Interconnect, allows for the establishment of enterprise-grade connections to Google Cloud Platform. This service offers two options: Dedicated Interconnect provides a private connection via a dedicated physical link, and Partner Interconnect allows connections through a supported service provider. Both options facilitate secure and high-throughput connections, ideal for high-volume data transfers and mission-critical applications requiring robust data privacy and low latency.

Multicloud Connectivity Strategies

Organizations employing a multicloud strategy use services from multiple cloud providers to optimize benefits such as reduced dependency on a single vendor, enhanced disaster recovery capabilities, and localized compliance with data residency requirements. Managing connectivity across these diverse platforms requires

Inter-cloud network routing: Sophisticated routing mechanisms that allow for seamless data flow between different cloud services

Cross-cloud data management: Tools that facilitate data integration and management across platforms, ensuring data consistency and accessibility

The selection and management of these connectivity technologies are important for creating a foundation for effective communication and coordination within sociotechnical systems. Whether through basic networking within a single facility, secure remote access via VPNs, or direct connections to cloud services, each technology plays a pivotal role in

CHAPTER 3 THEORY

ensuring that all components of a sociotechnical system can communicate effectively. As the complexity of organizational needs grows, especially with the adoption of cloud and multicloud environments, the strategic integration of these connectivity technologies becomes more critical, directly impacting the system's efficiency, scalability, and security.

The role of coordination in sociotechnical systems cannot be overstated—it is essential for ensuring that the intricate web of interactions within these systems is efficient, effective, and adaptable. With the rise of AI and other advanced technologies, we are on the cusp of a major transformation in how coordination is understood and implemented. This shift holds the potential not only to enhance existing systems but also to redefine what is possible in organizational management and operational efficiency, paving the way for more resilient and adaptable organizations.

Key Takeaways

This chapter has introduced four key theoretical concepts that are essential for understanding the unit-oriented approach to architecting complex systems. These concepts—Triadic Evolution, sociotechnical systems, Evolutionary Units, and coordination—have been presented in an approachable, nonscientific manner, making them accessible to readers from various backgrounds. Each concept builds upon the others to form a comprehensive framework that supports the effective design and management of intricate, adaptive systems.

CHAPTER 3   THEORY

Figure 3-12. Evolveburgh's mayor had a grand plan: to build the world's most advanced sociotechnical system to mine the universe's dark matter and dark energy. But the universe, with its infinite wisdom, had other ideas. After spending billions of dollars and countless iterations, the dark matter and dark energy that Evolveburgh mined on the first day turned out to be nothing other than dark chocolate and dark coffee. Disappointment quickly turned to delight as they switched to plan B. Now, Evolveburgh boasts cosmic chocolate mines and coffee drilling rigs, exporting 12 billion tons of "Euphorica" chocolate and 10 million barrels of "Eurikano" coffee a year to 173 countries. Who knew the answer to the universe wasn't 42, but a delicious 84% cocoa and a robust, full-bodied dark roast? So next time you're sipping on your morning coffee or enjoying a piece of dark chocolate, remember, it's not just a treat, it's a cosmic delicacy from the sweetest spot in the galaxy!

Here are the key takeaways from each of these foundational concepts:

- **Triadic Evolution**: Triadic Evolution underscores the interlinked development of biological, organizational, and technological progress. Understanding this concept is important because it highlights how human capabilities, social structures, and technological advancements synergistically evolve to drive the complexity and effectiveness of sociotechnical systems. This evolution sets the stage for building systems that are not only technologically advanced but also deeply integrated with human and social factors.

- **Sociotechnical systems**: The key takeaway from the concept of sociotechnical systems is the importance of viewing these systems as composed of sociotechnical entities—both integrative and operational—that work collaboratively to ensure the overall effectiveness and efficiency of the whole. This perspective highlights the need to design systems where the legitimate, purposeful, accountable, capable, and high-performing components are not only mutually supportive but also collectively contribute to achieving organizational goals in a seamless manner. Understanding this dynamic interplay is essential for creating systems that are not just functional but are optimized for both human interaction and technological efficacy.

CHAPTER 3 THEORY

- **Evolutionary Units**: Evolutionary Units are the building blocks of sociotechnical systems, characterized by their social cores and technological peripheries. The essential takeaway is that maintaining a strong social core while effectively leveraging technological extensions ensures that these units not only function optimally but also adhere to ethical guidelines and contribute positively to the system's overarching objectives. This balance is critical for the sustainability and ethical integrity of sociotechnical systems.

- **Coordination**: Coordination is identified as the mechanism that enables smooth and efficient interaction across different components of a sociotechnical system. The takeaway here is the shift toward more sophisticated coordination mechanisms, facilitated by advances in technology such as swarm intelligence and machine learning, which can dynamically enhance the interaction between system units. This evolution in coordination is essential for modern systems to be more responsive, adaptable, and capable of handling complex operational demands.

These takeaways encapsulate the fundamental concepts needed to appreciate and apply the unit-oriented approach in designing and managing complex sociotechnical systems, ensuring that they are robust, adaptive, and aligned with both human needs and technological advancements.

CHAPTER 4

Model

The author's expedition through the labyrinth of representations—spanning blueprints, models, metaphors, and even whimsical doodles—culminates in the selection of an essential metaphor for crafting sociotechnical systems. Balancing on the edge of cognitive exhaustion, he awards himself the "Abstraction Badge of Honor." This self-awarded token not only symbolizes his triumph in the realm of metaphorical mastery but also signifies his notable contribution to the field.

In Chapter 3, we distilled our journey into the realm of large sociotechnical systems into four key theoretical concepts. These concepts encapsulate the insights we've gained and serve as the key takeaways for our readers. They guide us toward a future where we can metaphorically model and illuminate large sociotechnical systems in a new light, providing a fresh perspective on their structure and function. Building on the concepts we've explored in the previous chapters, these concepts offer a road map for navigating the complexities of these systems.

Representations: From Mimicry to Modernity

From the earliest moments of human existence, the act of representation has been an intrinsic part of our identity. As children, we instinctively mimic the world around us, using toys to replicate our surroundings

CHAPTER 4 MODEL

and drawings to capture our perceptions. This innate inclination for representation distinguishes us from other species; our essence as humans is deeply rooted in our capacity to imitate, symbolize, and communicate through stand-ins of reality.

Figure 4-1. *Future architects at play: building dreams while grown-ups struggle to understand the blueprint. In Evolveburgh, kids are natural-born copycats, using toys to mirror the world in a fun-sized format and drawings to paint their vivid imaginations. This natural talent for imitation sets them apart from other species, especially the most enigmatic species of all—grown-ups!*

Historically, representation has held a central position in disciplines as diverse as aesthetics, semiotics, and, more recently, political theory. Since the dawn of the earliest civilizations, it has been the cornerstone of our understanding of art and symbols. Moving into the modern era, from the 18th century onward, representation assumed a pivotal role in political thought, shaping our perceptions of sovereignty, legislative authority, and the intricate relationship between the individual and the state, as proposed by thinkers like W.J.T. Mitchell.

CHAPTER 4   MODEL

The following illustrates the multifaceted nature of representation:

- Semiotic representation
  - **Stand for**: Symbols, signs, or images that represent ideas or concepts
- Political representation
  - **Act for**: Elected officials representing the interests of their constituents
- Social representation
  - **Speak for**: Advocates or activists representing the interests of a community
  - **Vouch for**: Endorsements or guarantees where one entity supports another
- Scientific representation
  - **Account for**: Theories or equations that explain observed phenomena
  - **Model for**: Models used to simulate or predict real-world phenomena
- Practical/functional representation
  - **Substitute for**: Direct replacements, like ingredients in cooking
  - **Serve as a proxy for**: Indirect measures or stand-ins for something else

Drawing from Aristotle's perspective on the matter, representations can be differentiated based on three defining aspects: the target they aim to depict, the form they adopt for the portrayal, and the medium they employ for this purpose. The "target" signifies what is being represented, the "form" delves into the configuration and style of representation, and

133

CHAPTER 4   MODEL

the "medium" relates to the tangible or intangible materials utilized in the representation process. While Aristotle referenced these as the "object," "manner," and "means," modernizing the terms to "target," "form," and "medium" provides a more intuitive understanding, especially for contemporary audiences and practitioners.

In the engineering realm, since the mid-19th century, drawings, schematics, blueprints, and specifications have served as media of representation, capturing the intricacies of systems, structures, processes, or phenomena. In numerous engineering contexts, such models are indispensable for comprehending and structuring phenomena.

While drawings, schematics, and blueprints indeed serve as effective means of representation, offering a distilled version of reality or a systematic depiction of an object or system, a more sophisticated medium—models—has risen to prominence. These are pivotal in exploring, elucidating, acquiring knowledge, forming concepts, crafting theories, and constructing other models.

Models: Beyond Blueprints to Constructing Realities

Historically, when one thought of representation in the realm of engineering or design, images of detailed drawings, meticulous schematics, or comprehensive blueprints might have come to mind. These traditional forms, which have been the backbone of design and construction for centuries, provide a clear snapshot of design, layout, and dimensions. They are invaluable in their own right, capturing the envisioned outcome with precision. Yet, they're predominantly static in nature. While they can illustrate how a system is supposed to look or be structured, they don't necessarily illuminate how it might behave or respond under varying conditions. In essence, while they offer a detailed picture, they don't always breathe life into that picture.

CHAPTER 4   MODEL

Enter the realm of modern modeling. In contemporary engineering and design contexts, the term "model" has evolved to encompass so much more than static representation. Today, when professionals speak of a "model," they often allude to dynamic tools capable of simulation or prediction. These can be mathematical constructs, computer simulations, or other interactive forms that not only represent a system but also allow for its exploration under diverse scenarios. Such models serve as powerful lenses, enabling us to examine, predict, and even manipulate the intricate dynamics of the target system.

Figure 4-2. In Evolveburgh, the City of Zillion Details, models aren't just doodles—they're the "Picassos" of the system world, painting a picture of complexity with a brush of simplicity. They transform the overwhelming into the "oh-I-see" and the incomprehensible into "isn't-that-a-piece-of-art." Here, a flowchart isn't just a flowchart; it's a treasure map leading to the "X" of understanding. A Venn diagram isn't just a Venn diagram; it's a Zen diagram, capturing the intersection of reality and dreams. And every bar graph is a rollercoaster ride of life's ups and downs that ends at the local bar, where every "bar" in the graph gets a "cheers!"

135

CHAPTER 4 MODEL

Models have transformed the way we approach complex systems, serving as bridges between mere representation and deeper insights. They distill the vast complexities of the world into formats that are not only comprehensible but also interactable. By crafting a model, we can capture the essential features of a system without being overwhelmed by every minute detail. This abstraction allows us to

- **Structure complex systems**: Models provide a scaffold, helping us to organize and understand the myriad components and interactions within larger systems.

- **Test hypotheses**: With models at our disposal, we can experiment, tweaking variables to explore various scenarios. This ability to manipulate models lets us validate or challenge our beliefs and assumptions about real-world systems.

- **Predict outcomes**: A well-constructed model serves as a crystal ball of sorts. By inputting current data, we can anticipate future events or behaviors, be it predicting the trajectory of a storm or forecasting economic growth.

- **Facilitate communication**: Especially in the realm of complex systems, a visual or interactive model can communicate ideas far more effectively than words alone. They can bridge the comprehension gap, making intricate concepts accessible to both experts and laypeople.

However, the true strength of a model lies in its accuracy. For it to be effective, a model must mirror the essential attributes of its target system. This means capturing the foundational mechanics, relationships, and variables that drive the system's behavior. As an example, a meteorological model aiming to predict a storm's path needs to incorporate factors like atmospheric pressure, temperature differentials, and prevailing wind currents.

CHAPTER 4   MODEL

To guide the complex process of modeling, we lean on a combined framework drawing from RIG Hughes's DDI and José A. Diez's Ensemble-Plus-Standing-For (EPS) accounts:

- **Denotation (what)**: At its core, a model must define what it represents. It should symbolize, replace, and point to its target, acting as a stand-in for the real-world system.

- **Existence (why)**: Beyond mere representation, a model must have a purpose. It should manifest the system's essence and provide rationale for its own existence, answering why it was created in the first place.

- **Directionality (where)**: Models are not isolated constructs. They originate from real-world observations and aim to provide insights back into the real world. This bidirectional relationship is vital for a model's relevance.

- **Demonstration (how)**: Once constructed, a model is not static. It's a tool for exploration. Through interaction and manipulation, we can probe its mechanics, structure, and underlying principles.

- **Interpretation (which)**: After experimentation, the insights gleaned from a model must be translated back to real-world implications. This step ensures that the findings are not only accurate but also actionable.

This framework isn't a linear progression but a road map for modelers. Beginning with denotation, a model captures the core of the system. As we progress, questions about its existence and its connection to the real world emerge. By the demonstration phase, the model is rigorously tested,

CHAPTER 4   MODEL

manipulated to gauge its responses. Finally, interpretation takes these findings, translating them into actionable insights about the real world.

The power of the framework lies in its holistic approach. Denotation establishes identity; existence guarantees relevance; directionality creates connections; demonstration ensures accuracy; and interpretation guarantees real-world applicability. Collectively, these stages ensure the model's adequacy, aligning it with intent and ensuring it meets standards and fulfills its purpose.

This comprehensive approach ensures that models are not just representations, but dynamic tools that enhance our understanding, drive innovation, and shape the future. As we delve deeper into the nuances of modeling and representation, it becomes evident that models, especially when bolstered by metaphors, have the power to transform our approach to large sociotechnical systems, offering gateways to understanding, prediction, and innovation.

Metaphors: Fitting the Most Meaning in the Least Space

The realm of modeling is multifaceted and intricate, and its various forms cater to different cognitive and explanatory needs. Drawing upon the insights of Thomas A. Sebeok and Marcel Danesi, we find that all phenomena of representation can be categorized into four overarching types. Each of these categories delineates a unique modeling activity, reflecting different ways in which we understand, represent, and interact with the world around us:

- **Singularized modeling (unitary):** This mode of modeling emphasizes a direct, one-to-one mapping between sign and meaning. An easily relatable example can be found in our daily interactions with traffic lights. Here, the color red universally and unequivocally signals "stop." The relationship between the sign (the red light) and its meaning (the instruction to stop) is straightforward, leaving little room for ambiguity.

- **Composite modeling (textual)**: As the name suggests, this approach derives meaning not from individual signs in isolation but from a sequence or combination of signs. A novel serves as a fitting analogy. While each word within it carries its own meaning, the true essence of the story emerges from the intricate arrangement and interplay of these words. The narrative is a composite, formed by piecing together individual linguistic signs in a particular sequence.

- **Cohesive modeling (code based)**: Rooted in predefined systems or codes, this form of modeling thrives on precision and operates within clearly set boundaries. Morse code exemplifies this type. Each letter of the alphabet is represented by a specific sequence of dots and dashes. The meaning of each sequence is fixed and is derived from the code's predefined system.

- **Connective modeling (metaphoric)**: Perhaps the most abstract of the four, connective modeling leverages analogies and metaphors to convey meaning. When one describes a startup's growth trajectory as "rocket-like," they aren't suggesting that the startup is an actual rocket. Instead, they're drawing on the familiar concept of a rocket's rapid ascent to capture the essence of the startup's swift growth. This approach connects two seemingly unrelated domains to shed light on one through the other.

CHAPTER 4 MODEL

Given humans' intrinsic cognitive tendency to relate new information to preexisting knowledge frameworks, metaphoric models stand out as particularly invaluable, especially when delving into the intricacies of large sociotechnical systems. By incorporating a familiar "source" alongside the model and target, these models bridge the gap between the known and the unfamiliar. For instance, in the metaphor "time is money," the tangible and familiar domain of finance aids in illuminating the abstract and often elusive concept of time. The relational structure embedded within metaphoric models makes them inherently more accessible and resonant.

Complex systems, regardless of their foundation—be it the perplexing realm of quantum physics, the multifaceted dynamics of biological processes, the vast expanse of advanced mathematical theories, or the interwoven web of large sociotechnical systems—pose significant challenges to comprehension. Traditional scientific models, while rigorous and detailed, often remain confined to the domain of specialists. Their precision, although admirable, can sometimes render them inaccessible to those not deeply entrenched in the discipline. Even for those well-versed in the field, the sheer complexity and multidimensionality of such systems can be overwhelming.

CHAPTER 4   MODEL

Figure 4-3. *In Evolveburgh, metaphors became a public epidemic, popping up everywhere, even in tax returns, building permits, utility bills, and probate records. The city had to enforce an unprecedented law—use more than three metaphors a day, and you're ticketed! They even hired a "Metaphor Monitor" to control the metaphorical chaos. One day, a man was caught using four metaphors in a single sentence at the grocery store and was promptly issued a metaphorical parking ticket. When he tried to pay it using slang metaphors, he was taken to court, faced metaphorical charges, and was sentenced to a month of literal language use. No more "raining cats and dogs" or "beating around the bush." He had to say exactly what he meant, which was a "hard pill to swallow." But in the end, he learned that, sometimes, "less is more," especially when it comes to metaphors!*

CHAPTER 4   MODEL

This is where metaphoric models come into their own, shining as beacons of understanding. For both specialists and general audiences, metaphors can serve as anchors, grounding abstract and challenging concepts in the realm of the familiar and tangible. Consider, for instance, the thought experiment of Schrödinger's cat in the domain of quantum mechanics. Here, the seemingly paradoxical notion of a cat that exists in a simultaneous state of being alive and dead offers a metaphorical entry point into the baffling world of quantum superpositions. While this metaphor doesn't capture every minute detail of the quantum state, it provides an intuitive starting point, a metaphorical foothold in a terrain that might otherwise seem impenetrable.

Figure 4-4. *Visitors to Evolveburgh quickly notice the residents' obsession with Schrödinger's cats, a special breed developed in the city for mysterious purposes. These cats not only outshine other cats in animal shows, but they also outperform dogs and even hamsters. They're so impressive; they occupy the top 100 spots... simultaneously!*

For the pioneering minds charting the unexplored territories of these complex systems, metaphors can also play an invaluable role as heuristic devices. By drawing parallels with known entities and familiar experiences, they can navigate the intricate corridors of their theories more effectively.

Metaphors can help in spotting patterns, making connections, and even generating novel insights that might remain obscured when viewed through a purely scientific lens.

But metaphoric models aren't just about simplification. Their true power lies in their ability to facilitate understanding without diluting the essence of what they represent. They weave together the new and groundbreaking with the familiar and understood, allowing even those not deeply versed in a field to grasp and appreciate the profound insights it offers.

Metaphors do more than merely describe or compare; they actively structure our understanding of the world. They are foundational cognitive models through which we interpret our experiences and make sense of new information. As eloquently articulated by Lakoff and Johnson, metaphors bridge reason and imagination. They provide a scaffold for our cognitive processes, merging the logical with the creative, thereby enabling a more holistic understanding.

In the domain of innovation, the role of metaphors cannot be understated. They are not mere rhetorical devices but powerful cognitive tools that shape how we perceive, interpret, and even reimagine the world. Metaphors facilitate the construction of more intricate cognitive models by building upon and extending existing ones. In situations where phenomena are beyond our direct experience—whether because they are too vast, too minuscule, too abstract, or simply too complex—metaphors offer a way in. They help us bridge our cognitive limitations, transforming the unseen into the seen and the incomprehensible into the comprehended.

When deploying metaphors, especially in the context of large sociotechnical systems, one must ensure their relevance, clarity, specificity, richness, and quality. An effective metaphor should be readily understood, offer multiple layers of insight, and represent its subject in a manner that is both comprehensive and illuminating. Achieving this balance ensures that the metaphor not only aids in understanding but also enhances it, paving the way for deeper insights and more meaningful interactions with the system at hand.

CHAPTER 4 MODEL

Throughout our exploration of large sociotechnical systems design, we have made deliberate and informed choices to best represent and understand these sophisticated systems. At the heart of our approach is the recognition that models serve as advanced forms of representation, offering a more dynamic and detailed lens compared to static blueprints or schematics. They provide a means to simulate, analyze, and predict the multifaceted interactions inherent in LSS. But even within the realm of models, we identified a spectrum of complexity and utility. At the pinnacle of this spectrum are metaphoric models, which we've chosen as our primary tools for LSS design. These models stand out due to their unique ability to bridge the known with the unknown, drawing from familiar experiences to elucidate complex systems. By leveraging the power of metaphor, we not only simplify intricate concepts but also retain their depth and richness, ensuring a comprehensive understanding. This fusion of advanced representation through models, coupled with the nuanced insights offered by metaphoric models, forms the bedrock of our approach to designing and understanding LSS.

As we transition to the next section, we'll delve into the process of selecting the most apt metaphor to aid in the design and understanding of the digital aspects of large sociotechnical systems.

The Journey: Designing the Quest for the Perfect Metaphor

When embarking on the challenging path of designing complex systems, architects often find themselves in need of a guiding principle, a thematic compass, if you will, that can bring clarity to the vast expanse of their vision. This guiding principle often takes the form of a metaphor. Much like an archaeologist meticulously brushes away layers of history to reveal a hidden artifact, the process of discovering the right metaphor is an exploratory journey filled with moments of wonder, introspection, and profound insights.

The challenge of designing a large sociotechnical system is unique in its own right. These systems are sprawling, comprising multiple facets and layers that interact in myriad ways. To encapsulate such a vast system within a singular, coherent metaphor demands a shift from conventional architectural paradigms. It requires architects to zoom out, take a bird's-eye view, and look at the LSS not just as a sum of its parts but as a holistic entity. And as they do so, they should remember that often the most powerful metaphors are derived from our everyday experiences and are deeply embedded in our linguistic expressions.

But how does one go about crafting such a metaphor? What kind of thought process and heuristics should architects employ to conceive a metaphorical image that stands as a true reflection of an LSS? It's not just about finding a neat comparison; the chosen metaphor should enhance comprehension, facilitate in-depth and structured conversations about the system, be easily integrated into daily discourse, and, more importantly, act as a catalyst for transformative thinking. It's worth noting that while there's value in dissecting an LSS to understand its individual components, doing so without keeping an eye on the overarching narrative can lead to a fragmented view.

Let's delve deeper into the systematic, three-phased approach to this metaphorical quest:

1. Problem Formulation: Immersion and Groundwork
   - The initial step involves defining the challenge in clear terms: "As an architect, I want to model a large sociotechnical system as a metaphor (M) to better communicate its complexities and interrelationships."

CHAPTER 4   MODEL

- This is followed by a deep dive into the knowledge acquisition phase. Here, architects can benefit from drawing insights from a rich repository of historical metaphors, studying ancient artifacts, and pondering over the underlying dynamics of the target system.
- The groundwork is equally important. It's about grounding the abstract in the tangible. This means observing real-life events or phenomena that echo the key characteristics of the LSS. Remember, the most resonant metaphors often have their roots in common, everyday experiences.

2. Metaphorical Imagination and Formation

- Here, the architect engages in brainstorming, ideating, and generating a plethora of potential metaphors. The goal is to anchor abstract concepts in concrete, universally understood terms.
- Each idea is then fleshed out, scrutinized, and refined. The metaphorical lens is adjusted and readjusted, ensuring that the perception of reality it offers is both clear and relevant.
- It's a balance of deductive reasoning, which is rooted in existing scholarly works and literature, and inductive reasoning, a more intuitive form of thought that leans heavily on imagination.

CHAPTER 4   MODEL

3. Evaluation: Testing the Metaphor's Aptness and Making the Final Choice

    - Each metaphor is put through a rigorous evaluation process. Does it truly bridge the gap between the known and the unknown? Does it resonate with the intended audience? Is it both architecturally robust and conceptually rich?

    - The chosen metaphor must not only be structurally sound but should also inspire fresh design perspectives. It should be versatile, promoting adaptability and flexibility in the face of evolving challenges and requirements.

In conclusion, the metaphorical journey, once embarked upon, reveals itself to be much more than a mere academic exercise. It's a transformative experience that reshapes the way architects perceive, understand, and approach their design challenges. The right metaphor serves as a beacon, illuminating the sophisticated maze of complexities, making the unfamiliar terrain of LSS seem both navigable and familiar.

As we stand at the threshold of the final part of our exploration, anticipation runs high. The subsequent section promises to unveil the metaphor that will be enshrined at the heart of our architectural framework, shaping perceptions, guiding design decisions, and offering a fresh perspective that will be instrumental in both conceptualizing challenges and devising solutions.

The Moment of Truth: Discovering the Metaphor for LSS

Identifying the right metaphor to LSS was a challenging and time-consuming endeavor. After sifting through numerous candidates, applying rigorous ranking and scoring mechanisms, and eliminating those that

CHAPTER 4   MODEL

did not measure up, six metaphors emerged that offered distinctive perspectives on conceptualizing LSS:

- Large Sociotechnical Systems is a City
- Large Sociotechnical Systems is a Hospital
- Large Sociotechnical Systems is a University
- Large Sociotechnical Systems is a Factory
- Large Sociotechnical Systems is an Army
- Large Sociotechnical Systems is a Fleet

To streamline our understanding and accentuate the digital facets of LSS, we adopted the "Digital Enterprise" as a representative model for LSS. This brought forth analogous metaphoric representations:

- The Digital Enterprise is a City
- The Digital Enterprise is a Hospital
- The Digital Enterprise is a University
- The Digital Enterprise is a Factory
- The Digital Enterprise is an Army
- The Digital Enterprise is a Fleet

Our evaluation of these metaphors was rooted in the foundational principle that "collectives"—emerging from the symbiosis of organizational evolution with technological extension—form the bedrock of enterprises. These collectives, representing human cohorts, retain full control over their technological facets and are intrinsically accountable for their actions.

Guiding our analysis was a set of criteria aimed at ensuring the metaphors were not just descriptive but also actionable and intuitive:

- **Relevant**: The metaphor must resonate, offering tangible references and rich descriptive prowess.

- **Well understood**: Universal comprehension is key. A metaphor loses its power if it is not easily grasped.

- **Rich**: Depth is fundamental. A metaphor that provides varied interpretations of a concept is invaluable.

- **Interesting**: It should spark creativity, leading to fresh design perspectives.

- **High quality**: The metaphor must portray the enterprise as an architectural marvel—coherent, elegant, adaptable, and sophisticated, seamlessly integrating its components.

Following our stringent evaluation process based on the aforementioned criteria, a discernible pattern emerged. The metaphors that consistently stood out as the most relevant, interesting, rich, and architecturally superior were invariably those rooted in large *physical* sociotechnical systems. These systems, whether they be ancient civilizations, sprawling cities, or meticulously coordinated fleets, have, over millennia, been refined through trial and error, innovation, and strategic design. The architectural brilliance of these physical systems is not just a testament to human ingenuity but also a rich repository of design principles, organizational strategies, and integration techniques. Delving into the annals of history, we uncover a treasure trove of insights: how ancient metropolises managed complex logistical challenges, how vast fleets ensured seamless coordination among diverse units, or how sprawling universities fostered innovation while maintaining order. As we venture into the digital age, designing large sociotechnical systems for the

CHAPTER 4   MODEL

future, these historical precedents offer invaluable lessons. They remind us that while the medium may change—from brick and mortar to bits and bytes—the foundational principles of design, organization, and integration remain consistent. Drawing from the wisdom of the past, we can sculpt the digital enterprises of the future, ensuring they are not only technologically advanced but also resilient, adaptable, and harmoniously structured.

The Digital Enterprise Is a City

The metaphor of a city, teeming with life, structure, and dynamic interactions, offers a rich imagery to understand the complexities and nuances of a Digital Enterprise. By drawing parallels between the multifaceted layers of urban landscapes and the integrated components of digital enterprises, we can derive insights into their design, functionality, and evolution.

> **Relevant**: A well-planned city, with its harmonious interplay of various districts, infrastructure, and inhabitants, serves as a compelling parallel to a Digital Enterprise. Much like how different parts of a city serve distinct functions—commercial districts, residential areas, industrial zones—a Digital Enterprise comprises varied systems and departments, each catering to specific needs. Not all aspects, however, align seamlessly with the realm of a Digital Enterprise. The vastness and intricacy of a city can indeed become overwhelming when directly paralleled with digital architectures. The multilayered components of a city, from its road networks to parks and public spaces, might not find a direct correlation in the digital world, potentially leading to ambiguities or over-complications in design.

CHAPTER 4   MODEL

Well understood: The concept of a city—its layout, dynamics, growth, and challenges—is universally understood. Almost everyone has experienced life in a city or is at least familiar with its structures and operations. By likening a Digital Enterprise to a city, architects and designers tap into a universally relatable metaphor, simplifying complex digital concepts and making them more accessible to stakeholders.

Rich: A city's richness in terms of its structures, dynamics, and functions can be both an asset and a hindrance. While it offers a plethora of design inspirations—the growth, expansion, or even decline of cities can offer insights into scalability, adaptability, and sustainability in the digital realm—the risk lies in getting bogged down by the metaphor's intricacies. Designers might find themselves entangled in trying to map every city element to a digital component, which could be counterproductive.

Interesting: Cities are undeniably fascinating entities—hubs of innovation, culture, and diversity. They face myriad challenges, from infrastructure to sustainability, and constantly seek innovative solutions. This aspect of cities can inspire novel problem-solving approaches in Digital Enterprise design. The multifaceted nature of cities—where commerce, art, technology, and more coexist—can drive holistic, integrative thinking in digital architecture.

CHAPTER 4   MODEL

> **High quality**: Cities, especially well-planned ones, exemplify harmony between form and function. The blend of aesthetic appeal with utility, the balance between green spaces and urban sprawl, and the emphasis on sustainable growth can all serve as guiding principles for designing high-quality digital enterprises. However, translating the balance and harmony into the digital realm is challenging, because "decorative" or "recreational" elements seen in well-planned cities—pedestrian bridges, underground passages, viaducts, public squares, parks, and water areas—can consume resources and complicate the digital system without tangible benefits. The challenge is to distill the essence of a city's high-quality design while avoiding the trap of unnecessary embellishments.

Taking everything into account, the "Digital Enterprise Is a City" metaphor offers a multidimensional framework for conceptualizing and designing digital systems. It encapsulates the essence of diversity, adaptability, and harmonious growth, urging architects and designers to envision digital enterprises that are dynamic, integrative, and user-centric, much like thriving urban landscapes. On the other hand, while the city metaphor brings forth a volcano of ideas, it also comes with potential pitfalls. Its complexity and vastness might indeed be overwhelming, leading architects and designers astray with aspects that are more distractive than constructive. Care must be taken to harness the metaphor's strengths while being wary of its complications, ensuring that the resultant digital design is streamlined, functional, and devoid of unnecessary complexities.

CHAPTER 4   MODEL

The Digital Enterprise Is a Hospital

In the clean and quiet corridors of a hospital, where specialized teams and cutting-edge equipment converge to heal and innovate, we find a mirror to the Digital Enterprise. The hospital metaphor emphasizes the balance of expertise, collaboration, and the seamless integration of people and technology.

> **Relevant**: Hospitals, with their specialized departments and teams, serve as a striking parallel to Digital Enterprises. Each department, be it cardiology, neurology, or radiology, represents a collective of professionals, harmoniously extended by dedicated physical spaces and specialized equipment. In a Digital Enterprise, this can be likened to specialized teams or departments, each extended and enhanced by tailored digital tools and platforms to meet their unique objectives. The core of this metaphor revolves around the seamless integration of social structures (teams) with their extensions (equipment, tools, and furniture).
>
> **Well understood**: The hospital model is universally recognized. People understand that a radiology team, for instance, is complemented by specific equipment and spaces like X-ray machines and MRI suites. This offers a tangible model for stakeholders to grasp how specialized teams in a Digital Enterprise can be similarly extended and empowered by specific digital structures.

153

Rich: The hospital metaphor is multifaceted. From emergency units that require rapid response tools and spaces to research wings needing advanced equipment and quiet labs, each department's unique needs are met with tailored physical extensions. This richness provides a blueprint for digital architects, showing how specialized teams can be paired with bespoke digital tools to maximize efficiency and output.

Interesting: The dynamic between hospital teams and their equipment is a dance of precision, trust, and synergy. This intricate relationship can inspire a similar synergy in Digital Enterprises, emphasizing the importance of aligning digital tools with team needs, ensuring neither is redundant or overpowering the other.

High quality: A hospital's commitment to patient care is manifested in how its teams are perfectly paired with their physical spaces and tools. An operating theater, for instance, is the epitome of design precision, ensuring surgeons have everything they need within arm's reach. This level of meticulous design can serve as a gold standard for Digital Enterprises, driving home the need for harmonious integration between teams and their digital extensions.

In conclusion, the "Digital Enterprise is a Hospital" metaphor underscores the importance of harmoniously extending specialized teams with tailored tools and platforms. Just as a hospital's efficacy hinges on the symbiotic relationship between its teams and their equipment, a Digital Enterprise's success is rooted in the seamless integration of its teams with their digital counterparts. This metaphor serves as a guiding light, emphasizing synergy, precision, and tailored design in the realm of digital architecture.

The Digital Enterprise Is a University

The hallowed halls of a university, a bastion of knowledge and collaboration, offer profound insights into the structure of a Digital Enterprise. Drawing parallels with academic departments and research wings, we explore how specialization and collective growth shape the digital realm.

> **Relevant**: Universities are renowned for their rich traditions, systems, and structures, underpinning the pursuit of knowledge. When an architect or designer considers the Digital Enterprise in this light, they might imagine a system where diverse departments or faculties function autonomously yet collaborate for a collective goal and the physical infrastructure, like buildings or labs, represents their extensions. Just as universities harbor a myriad of subjects, approaches, and methodologies, so too can an LSS encompass a vast array of functionalities, all bound by a shared mission.
>
> **Well understood**: The university metaphor is universally recognized. Most people are familiar with the roles, hierarchies, processes, and dynamics of a university setting. By likening the Digital Enterprise to such a well-known entity, architects and designers can easily tap into a shared understanding, enabling smoother communication and visualization of the system's blueprint.

Rich: Universities offers depth numerous angles for exploration, from overarching faculties to niche research groups. This richness allows for multiple interpretations and design pathways, ensuring the metaphor remains versatile and adaptable. However, when juxtaposed against the backdrop of a city with its infrastructural and socio-cultural dynamics, the richness can be overwhelming. Designers might grapple with how to map every university facet, from student unions to public lectures, onto a digital enterprise framework.

Interesting: Universities are hotbeds for innovation, research, and development. By viewing the Digital Enterprise as a university, designers are inspired to push boundaries and explore novel solutions, much like researchers or students on a campus. This perspective can lead to breakthroughs, encouraging a culture of continuous learning and innovation within the LSS.

High quality: Universities exemplify architectural prowess—both in their physical structures and organizational designs. They balance autonomy (departments running their courses) with cohesion (a unified degree system). In a similar vein, the Digital Enterprise should be designed as a cohesive yet flexible entity. The metaphor suggests an elegant integration of parts, with each "department" or module seamlessly interlinking. Furthermore, like universities which evolve their courses based on societal needs and advancements, the Digital Enterprise should be elastic, reconfigurable, and

CHAPTER 4   MODEL

transformative. This metaphor not only helps identify the key components of the system but also provides insights into how they might interact, adapt, and evolve over time.

In summary, the "Digital Enterprise is a University" metaphor encapsulates the essence of a dynamic, evolving, and interconnected system. It prompts architects and designers to think of LSS as entities that are both rooted in tradition yet ever-evolving, balancing autonomy with integration, and fostering a culture of continuous learning and innovation. However, adopting this metaphor comes with the risk of importing complexities associated with a city. Universities, with their campus roads, recreational areas, and communal spaces, introduce elements that might not have direct parallels in a Digital Enterprise, potentially leading to design ambiguities. It's essential to extract the core strengths of the university model—its specialization, structured approach, and emphasis on research and innovation—while being cautious of importing broader urban complexities that might not be directly relevant to digital enterprise design.

The Digital Enterprise Is a Factory

The rhythmic cadence of a factory, with its precision, efficiency, and division of labor, resonates with the mechanics of a Digital Enterprise. This metaphor underscores the importance of streamlined processes, innovation, and the harmonious interplay between diverse units. On the other hand, factories, with their logistical routes, supply chains, and integration with urban infrastructure, bring in elements that may not have straightforward parallels in a Digital Enterprise. This could introduce design ambiguities or distractions.

CHAPTER 4   MODEL

Relevant: Factories are structured environments dedicated to producing goods efficiently and consistently. In this light, a Digital Enterprise can be visualized as a factory where digital products or services are crafted. Just as factories have different production lines, machines, and processes, a Digital Enterprise comprises varied systems, tools, and workflows, each tailored for specific tasks. The focus on optimization, efficiency, and output quality in a factory can resonate with the goals of many digital enterprises.

Well understood: The factory model is universally recognized. Its emphasis on production, standardization, and efficiency is something most people are familiar with, either directly or indirectly. This makes the metaphor a potentially effective communication tool, providing a common ground for architects, designers, and stakeholders to discuss and visualize the structure and dynamics of a digital enterprise.

Rich: Factories offer a multilayered perspective. There's the macro view, where one looks at the factory's overall production capacity, strategies, and output. Then there's a micro view, diving into individual production lines, machinery specifics, and quality control measures. This granularity can be paralleled in digital enterprises, from overarching business strategies to specific digital solutions and tools.

Interesting: The evolution of factories, from the days of the industrial revolution to modern automated setups, can inspire a trajectory of growth and innovation in the digital realm. Concepts like lean manufacturing, which emphasize waste reduction and process optimization, can stimulate analogous strategies for digital enterprise efficiency and optimization.

High quality: Factories epitomize precision, consistency, and quality control. Every component, from raw materials to machinery, plays a key role in the final product's quality. Drawing from this, the Digital Enterprise should prioritize meticulous design, rigorous quality checks, and a relentless pursuit of operational excellence. The metaphor underscores the need for a harmonious integration of various components, ensuring that each part, no matter how small, aligns with the enterprise's quality standards.

On the whole, the "Digital Enterprise is a Factory" metaphor encapsulates a vision of precision, efficiency, and quality. It guides architects and designers toward conceptualizing digital enterprises that are structured yet flexible, consistent in output, and continuously optimized for peak performance. However, when one views the factory within the broader context of a city, complexities arise. Factories, with their logistical routes, supply chains, and integration with urban infrastructure, bring in elements that may not have straightforward parallels in a Digital Enterprise. This could introduce design ambiguities or distractions.

CHAPTER 4 MODEL

The Digital Enterprise Is an Army

The disciplined formations of an army, marked by strategy, hierarchy, and adaptability, provide a unique lens to view the Digital Enterprise. Emphasizing coordination, structure, and dynamic response, the army metaphor offers lessons in resilience and strategic planning.

>**Relevant**: An army's structured approach, from specialized units to their strategic bases like headquarters or barracks, provides an interesting parallel for a Digital Enterprise. While the units represent specialized teams, the bases can be seen as the foundational platforms or environments that support these teams. Each unit, be it infantry, artillery, or intelligence, represents a collective of personnel extended and augmented by specific equipment, tools, and tactics. In the context of a Digital Enterprise, this can be likened to specialized teams or departments, each being enhanced and empowered by specific digital tools and platforms. The metaphor emphasizes the alignment of specialized teams with the right resources to optimize their function and impact. However, drawing a direct correlation might be complex given the unique dynamics and functions of military bases compared to digital operational platforms.
>
>**Well understood**: The structure and organization of an army, with its clear hierarchies, divisions, and strategies, are generally recognized. However, the intricate details of military operations, strategies, and inner workings might not be universally understood. Furthermore, for individuals with

pacifist beliefs or those who have reservations about militaristic undertones, this metaphor might not resonate or even be met with resistance. It's important to be cognizant of these perspectives when presenting the metaphor to a diverse audience.

Rich: The army metaphor, with its multilayered integration of units, equipment, and bases, offers a depth of insight. Each base, whether it's a secluded barracks or a fortified headquarters, is tailored to support specific units and operations. This richness presents an expansive framework for digital architects, showcasing how specialized teams within an enterprise can be paired with both bespoke digital tools and dedicated operational environments to ensure cohesion and efficiency.

Interesting: While the disciplined formations and strategic maneuvers of an army can inspire innovative approaches in Digital Enterprise design, not all stakeholders might find this metaphor compelling. For some, especially those with strong pacifist inclinations, the very notion of using a military analogy might be off-putting or even objectionable. While it can offer fresh design perspectives to some, it might alienate others, making it a potentially polarizing choice.

High quality: An army's meticulous alignment of its units with their respective equipment and bases showcases strategic planning and precision. This can serve as a benchmark for Digital Enterprises. However, translating this to the digital realm

CHAPTER 4   MODEL

>requires discernment, ensuring that the essence of the metaphor is captured without getting entangled in military specifics that might not be directly applicable.

Overall, it may be said that while the "Digital Enterprise is an Army" metaphor provides deep insights into coordination, structure, and strategic alignment, it should be employed judiciously, particularly given its potential sensitivity among certain segments of the audience. The metaphor should serve as a catalyst for thought and exploration, not a rigid blueprint. It's paramount that the final digital architecture, while drawing inspiration from this metaphor, remains versatile and tailored to the unique aspirations and values of the enterprise.

The Digital Enterprise Is a Fleet

The vast expanse of the sea, navigated by a coordinated fleet, mirrors the expansive digital landscape of an enterprise. The fleet metaphor, with its emphasis on modularity, interoperability, and strategic movement, captures the essence of a Digital Enterprise navigating the ever-evolving technological waters.

>**Relevant**: Fleets, with their coordinated collections of vessels, each serving a unique purpose, are an apt analogy for Digital Enterprises. Just as each ship, aircraft, or spacecraft in a fleet is a specialized entity equipped for specific tasks, teams within a Digital Enterprise can be seen as distinct units, each extended and empowered by tailored digital tools and platforms. The metaphor emphasizes the idea of a harmonious collective, where each entity, while unique, contributes to the overarching mission of the fleet—or in this case, the enterprise.

Well understood: The basic structure and operation of a fleet—be it naval, aerial, or even space oriented—are generally graspable. Stakeholders can easily understand that a battleship, for instance, serves a different purpose and is equipped differently than a supply ship or a reconnaissance aircraft. This provides a tangible analogy for how specialized teams in a Digital Enterprise should be complemented by bespoke digital platforms and tools.

Rich: The fleet metaphor offers multifaceted insights. From the strategic positioning of an armada to the nuanced roles of each ship within a flotilla, the metaphor provides a wealth of scenarios and dynamics. This richness can serve as a comprehensive blueprint for digital architects, illuminating how specialized teams within an enterprise can be harmonized with specific digital tools to ensure cohesion and optimal performance.

Interesting: The dynamics of fleet operations, from coordinated naval maneuvers to synchronized aerial formations, can inspire innovative approaches in Digital Enterprise design. The metaphor encourages a vision of a Digital Enterprise as a synchronized entity, where collaboration, strategy, and precision drive the collective forward.

CHAPTER 4   MODEL

High quality: Fleets epitomize precision, coordination, and strategic excellence. Each vessel's design and equipment are meticulously tailored to its role within the fleet. This commitment to specialization and excellence can serve as a benchmark for Digital Enterprises. It emphasizes the seamless integration between teams and their digital extensions, ensuring operational precision and excellence.

Figure 4-5. In Evolveburgh, a survey on digital enterprise metaphors went off course when the local pirate-themed amusement park's influence steered residents toward a fleet metaphor

In general, it can be said that the "Digital Enterprise is a Fleet" metaphor emerges as a compelling and versatile tool. It underscores the importance of harmonious integration, specialization, and strategic coordination. Serving as both a guiding light and an inspirational source, the metaphor can effectively shape the vision and design of Digital Enterprises, making it a robust choice for architects and designers.

CHAPTER 4   MODEL

The Final Verdict: The Digital Enterprise Is a Fleet

Upon reflection on the various metaphors, several patterns and nuances surfaced. The Hospital, University, and Factory metaphors, while providing valuable perspectives on LSS design, echo some of the complexities inherent in the City metaphor. These metaphors, though rich in detail, might introduce layers that could potentially divert or overwhelm architects and designers, making the design process more intricate than needed. Meanwhile, the Army metaphor, with its structured and strategic undertones, indeed offers depth. Yet, it's not without challenges. Its detailed intricacies might not resonate universally, and its militaristic nuances could be contentious, especially for those with pacifist leanings or reservations about such analogies.

However, shining distinctively among these is the Fleet metaphor. Its architectural elegance, combined with its streamlined complexity, makes it a standout. The Fleet metaphor encapsulates versatility and clarity without the encumbrance of unnecessary complications. It offers a shared vision, enriches the design vocabulary, and illuminates novel pathways for LSS design. In its neat packaging and lean structure, the Fleet metaphor emerges as the prime choice for metaphorically modeling large sociotechnical systems.

Key Takeaways

This chapter delved into the profound impact of models on understanding and designing complex sociotechnical systems. From the fundamental acts of representation inherent in human nature to the sophisticated employment of models in modern system architecture, this chapter explored the essential role of models as tools for simplification, comprehension, and communication. Here are the key takeaways from this exploration:

CHAPTER 4   MODEL

- **The human essence of representation:**
  Representation is a core aspect of human identity, evident from our earliest interactions with the world. This chapter highlights how, from childhood, humans use toys, drawings, and later more structured forms like schematics and blueprints, to understand and interact with their surroundings. This innate inclination for representation is what sets humans apart and forms the foundation for the development of more complex models.

- **The evolution and role of models**: Models are not just representations but are tools for understanding, conceptualizing, and interacting with reality. They distill complex realities into manageable forms, allowing us to grasp what is otherwise too broad, too deep, or too complex to understand directly. By capturing the essential features of systems, models enable us to engage with, modify, and improve these systems without getting lost in unnecessary details.

- **Accessibility and comprehension challenges**: Traditional scientific models, while valuable for their precision and detail, often remain limited to domain specialists due to their complexity. This chapter underscores the importance of metaphoric models in making sophisticated concepts accessible and relatable to a broader audience. Metaphors help bridge the gap between complex scientific understanding and general comprehension, making them particularly valuable in the field of sociotechnical system design.

CHAPTER 4 MODEL

- **Metaphoric models as tools for design**: The "Large Sociotechnical System is a Fleet" metaphor is introduced as a powerful tool for conceptualizing and designing modern enterprises. This metaphor highlights the importance of structure, coordination, specialization, and integration—key attributes for any large-scale system striving for efficiency and effectiveness. Through this metaphor, the chapter illustrates how abstract concepts can be grounded in familiar terms, thereby enhancing both the design process and the communicability of complex system architectures.

- **Influence on system architecture**: Metaphoric models like the fleet metaphor do more than simplify complex ideas; they also inspire and shape the actual architecture of sociotechnical systems. By providing a coherent and relatable framework, such models guide system architects and designers in structuring components and processes in a way that aligns with the metaphor's implied principles and dynamics.

This chapter reveals how models, especially metaphoric ones, are indispensable in the realm of sociotechnical systems, offering not just a means of representation but a framework for deeper understanding and practical application in system design. These models serve as indispensable instruments for architects and designers, helping them to navigate the complexities of large systems and to communicate their visions effectively.

CHAPTER 5

Architecture

> *The author, with understated grace, introduces Unit Oriented Enterprise Architecture—a revolutionary method for architecting large sociotechnical designs, which serendipitously sprang from his own genius. Transitioning from the quiet achiever to the vanguard of his own design philosophy, he peppers his text with perceptive idioms, "strategic pivot" resurfacing with subtle finesse. If "synergy" once stood as the zenith of business jargon, prepare for an ascent into the stratosphere of buzzword mastery.*

In the ever-evolving landscape of technology and organizational design, the quest to find a guiding metaphor for large sociotechnical systems culminated in the adoption of the Fleet. This metaphor, as explored in the previous chapter, stood out not just for its architectural elegance, but for its ability to encapsulate the complexities, dynamics, and intricacies inherent to LSS. A fleet—comprising a harmonious blend of individual ships, each a marvel in its own right but achieving its fullest potential when operating as part of a coordinated whole—offers a compelling parallel to modern digital enterprises.

As we embark on this chapter, we'll delve deeper into the Fleet metaphor, unpacking its various facets and drawing parallels to the design and architecture of LSS. We aim to not just understand the Fleet as a metaphor but to translate its principles into actionable insights for architecting robust, adaptable, and efficient sociotechnical systems. By drawing from the vast expanse of maritime knowledge and aligning it with

CHAPTER 5   ARCHITECTURE

the nuances of digital enterprise design, we set sail on a journey to chart the uncharted, navigate challenges, and anchor our understanding in a blueprint for future LSS endeavors.

Architectural Approach

The endeavor to architect large sociotechnical systems is deeply influenced by the philosophy and attributes integral to the Fleet metaphor. Central to this architectural strategy are two pivotal concepts: the comprehensibility of the architecture itself and the changeability of the LSS as a whole. This approach aims to embed these important qualities deeply within the architecture of LSS, ensuring that the systems are not only adaptable and dynamic but also clear and understandable in their design and structure.

Changeability, in the context of LSS, reflects the system's inherent capacity to evolve and adapt. It's about crafting an architecture that's not just robust and resilient but also flexible and adaptable, capable of responding to new challenges and opportunities with ease. This dynamism is akin to a fleet that must navigate varying sea conditions and missions, requiring an architecture that supports scalability, adaptability, transformability, and reconfigurability.

Comprehensibility, on the other hand, is about ensuring that the architecture remains clear, legible, and accessible. The goal is to make the system's structure and operations transparent and understandable to all stakeholders, much like the clear, discernible structure and roles within a fleet. This clarity aids in managing and navigating the complex interplays within a large digital enterprise, enhancing overall effectiveness and efficiency.

In shaping the architecture of LSS, we embrace a simple methodology that comprises three essential processes: selecting components, positioning components, and relating components. Each process plays a critical role in constructing a cohesive, efficient, and adaptable system.

CHAPTER 5   ARCHITECTURE

Selecting components involves choosing the right elements that form the core of the LSS, like selecting ships for a fleet, each serving a specific purpose and function. Positioning components is about strategically placing these elements within the system's architecture to optimize performance and interaction, much like forming a fleet into a battle-ready formation. Finally, relating components is about defining the interaction and communication between these elements, ensuring that they work together harmoniously and effectively, similar to the coordination required among ships in a naval fleet. Together, these processes form the backbone of a robust and responsive LSS architecture.

- **Selecting components**: The first step in the architectural process is like choosing the right vessels for a fleet. Each component—whether a submarine, polar research vessel, or spaceship—should be selected not only for its individual purpose but also for how it contributes to the creation of a cohesive fleet. This selection is guided by the mission, strategy, and operational context, ensuring that each vessel adapts to changes while enhancing the overall clarity and purpose of the fleet.

CHAPTER 5   ARCHITECTURE

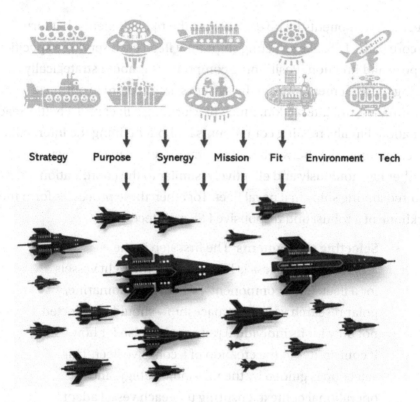

Figure 5-1. *This diagram illustrates the selected components for a space fleet, shown at the bottom, which have been chosen based on key decision factors—purpose, mission, environment, technology, fit, synergy, and strategy. At the top are the components that were not selected, represented in grayscale, indicating they were considered but did not fit the overall mission and strategy of the fleet. The fleet represents the optimal choice that best aligns with the system's goals and environmental requirements*

- **Positioning components**: Once selected, the next critical step is positioning these components within the system's architecture. This positioning is strategic, much like the formation of ships in a fleet. Each component must be placed in a way that maximizes its effectiveness, contributes to the system's resilience, and maintains harmony and balance within the overall architecture. This involves considering factors such as interdependencies, proximity to product and labor markets, and operational efficiency.

Figure 5-2. *Strategically positioned enterprise components: This image represents an enterprise modeled as a fleet of flying saucers, with key components strategically deployed across a planet resembling Earth. The visual metaphor highlights the dynamic and interconnected nature of modern enterprise architecture, where each component plays a key role in ensuring global reach and contributing to the overall success of the system*

- **Relating components:** The final piece of the architectural triad is defining the relationships between components. This phase involves establishing the protocols, interfaces, and communication channels that bind the components into a cohesive system. Like ships in a fleet that must operate in concert, the components of an LSS must be connected yet able to function independently. This interplay of relationships is key to ensuring that the system can withstand and adapt to both internal shifts and external pressures.

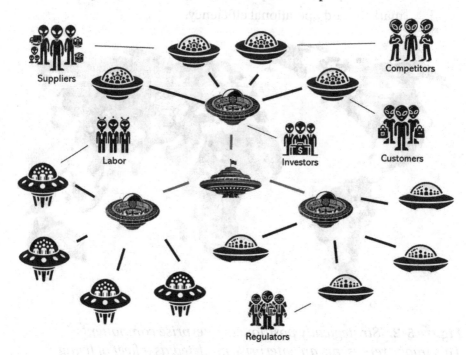

Figure 5-3. *The relationships within a sociotechnical system on an Earth-like planet and with external entities (suppliers, labor, customers, regulators, investors). The arrows depict interactions between these components, showing how the system depends on external inputs and is influenced by factors like competition and regulation*

In summary, the architectural approach to LSS, inspired by the Fleet metaphor, is about creating a system that balances changeability and comprehensibility. By carefully selecting, positioning, and relating components, we can craft an architecture that not only meets the immediate needs of the enterprise but is also poised to adapt and evolve in the face of future challenges and opportunities.

CHAPTER 5   ARCHITECTURE

Component Selection Patterns of LSS Architecture

In the architectural design of large sociotechnical systems, the selection of components is a foundational decision, shaping the system's overall character and functionality. This selection isn't limited to choosing uniform entities, like ships in a fleet, but extends to diverse types, including various technologies, platforms, and operational methodologies. We categorize these selection strategies into three types: homogeneous, heterogeneous, and hybrid, each carrying its implications for the system's comprehensibility, adaptability, and overall architecture.

Homogeneous Selection

Homogeneous selection emphasizes uniformity and consistency across all components, not just in their functionalities but also in their meaning, design, and implementation. This uniform approach ensures that each component is closely aligned in its core characteristics and operational dynamics. Homogeneous selection also promotes efficiency and predictability in system performance, as uniformity reduces the complexity of interactions between components. Moreover, it fosters a harmonious integration of parts, leading to a more robust and resilient system overall. This approach can be particularly beneficial in fields, where consistency and coordination are key to success.

On the positive side:

- **Comprehensibility**: The similarity among components makes the system easier to understand and navigate, streamlining modeling notation and pattern recognition.

CHAPTER 5   ARCHITECTURE

- **Simplicity in integration**: Homogeneity simplifies integration and management, providing a clear and consistent operational approach.

However, there are some drawbacks:

- **Limited flexibility**: Uniform components may not cater to specific or unique requirements, limiting the system's ability to adapt to diverse needs.
- **Systemic vulnerability**: A failure in one component might mirror across similar components, posing risks of widespread system issues.

Heterogeneous Selection

The heterogeneous approach values diversity, incorporating a variety of components that differ not only in their roles and functionalities but also in their logical, physical, and digital structures. This mix of distinct types allows each component to be specifically tailored for unique tasks, addressing a broad spectrum of operational needs within the LSS.

On the upside:

- **Resilience and adaptability**: The variety in components enhances the system's ability to handle diverse tasks and adapt to changing requirements.
- **Specialized optimization**: Components can be optimized for particular functions, improving performance in certain areas of the system.

On the downside, however:

- **Reduced comprehensibility**: The diversity can complicate understanding and managing the system, with increased complexity in modeling and pattern identification.

- **Integration challenges**: Coordinating and integrating a wide range of components requires more sophisticated approaches, potentially leading to higher complexity and cost.

Hybrid Selection

Combining the homogeneous and heterogeneous approaches, the hybrid selection aims to strike a balance between uniformity and diversity.

Looking at the advantages:

- **Flexibility and balance**: The hybrid approach offers flexibility, allowing for tailored responses to different operational scenarios, balancing predictability with adaptability.
- **Strategic optimization**: It enables strategic placement of heterogeneous elements within a homogeneous framework, optimizing performance while maintaining overall coherence.

Considering the disadvantages:

- **Complex decision-making and integration**: Determining the right mix and integrating heterogeneous components into a homogeneous base can be challenging.
- **Potential compromise in comprehensibility**: While offering balance, the hybrid approach might still reduce comprehensibility compared to a purely homogeneous structure, requiring careful planning and clear communication to mitigate complexity.

CHAPTER 5   ARCHITECTURE

In designing LSS architectures, the choice between homogeneous, heterogeneous, and hybrid selection patterns profoundly influences not just the system's functionality and efficiency but also its ease of understanding, modeling, and adaptation. These choices reflect a trade-off between the ease of comprehensibility and the flexibility to adapt and specialize, underlining the importance of strategic planning and clear architectural vision.

Favoring Uniformity in High-Level Selection Patterns

In crafting large sociotechnical systems, a strategic approach that promotes uniformity and consistency in the architectural elements, particularly at the higher levels of the system's structure, is essential. This approach is not only about ensuring operational efficiency but also about maintaining clarity and simplicity in the system's architecture. By adhering to a set of fundamental principles at these upper levels, we can build LSS that are not only effective but also comprehensible and harmonious in design.

The strategic choice to focus on uniformity and consistency in selecting architectural components aims to achieve the following:

- **Capability to build effective LSS**: The primary goal at the top levels is to establish a framework capable of supporting the creation and operation of an effective LSS. This means selecting architectural elements that contribute to a strong, resilient, and efficient system.

- **Commanding comprehensibility and clarity**: It's essential that the architecture at these levels commands comprehensibility and clarity. This clarity is achieved through legible, well-defined structures and interfaces that are easy to understand and navigate, making the system more accessible to architects, designers, and stakeholders.

- **Conveying clear and simple meaning:** The elements used should convey a clear and simple meaning, avoiding unnecessary complexity or ambiguity. This simplicity in meaning aids in ensuring that everyone involved has a consistent understanding of the system's architecture and its components.

- **Simplifying the work of LSS architects:** By maintaining uniformity and consistency, the architecture simplifies the decision-making and design processes for LSS architects. This uniformity allows architects to apply similar principles and patterns across different parts of the system, reducing complexity in architectural planning and implementation.

- **Fostering effective communication:** A coherent and uniform top-level architecture fosters effective and unambiguous communication among all participants involved in the LSS. It ensures that messages and intentions are clearly transmitted and understood, minimizing the risks of misinterpretation and errors in the system's development and operation.

Adopting a uniform and consistent approach in the top two levels of LSS architecture aligns with our broader architectural solutions, enhancing the system's overall effectiveness and integrity. By focusing on these elementary requirements, we ensure that the LSS not only functions efficiently but also remains manageable, comprehensible, and coherent, laying a solid foundation for the subsequent layers of more diverse and specialized architectural elements.

CHAPTER 5   ARCHITECTURE

Figure 5-4. *Evolveburgh's mayor almost clinched a Nobel Prize in Standardization, until the Committee realized that the city's "uniformity" translated to a social core extended by a technological periphery...a definition so broad, it could mean just about anything*

Structural Patterns of LSS Architecture

The choice of structural pattern plays a pivotal role in defining how components interact, communicate, and function as a part of the overall system. The structural patterns—Bus, Ring, Star, Tree, Mesh, and various Hybrids—each offer distinct advantages and challenges. Understanding these can help in selecting the most appropriate pattern for specific

LSS needs. In addition, the selection of a structural pattern significantly influences the scalability and adaptability of the system. A well-chosen pattern can facilitate growth and changes in requirements, thereby ensuring the system remains effective and relevant over time.

Bus

In a Bus structure, all components are connected along a central axis or backbone, similar to nodes along a linear path.

- **Pros**: Simplicity in setup and expansion; straightforward for adding or removing components.
- **Cons**: Limited flexibility; heavy reliance on the central backbone can lead to bottlenecks in operations and management.

Ring

A Ring structure connects components in a circular sequence, where each component directly interacts with two adjacent components.

- **Pros**: Equitable distribution of roles and responsibilities; inherently collaborative and resilient to single points of failure.
- **Cons**: Can face challenges in coordination and consistency; disruptions in one segment can affect the entire loop.

Star

In the Star pattern, components radiate out from a central core or hub.

- **Pros**: Centralized management and control; efficient in terms of resource allocation and decision-making.
- **Cons**: Overreliance on the central hub can lead to vulnerability and inefficiencies if the hub is compromised or overloaded.

Tree

This hierarchical structure features a root node with multiple levels of branching nodes.

- **Pros**: Clarity in role delineation and authority; scalable and manageable in a top-down approach.
- **Cons**: Rigidity in structure; dependency on the root and higher-level nodes can create bottlenecks and limit flexibility.

Mesh

Mesh structures feature a network where components are interconnected, often in a nonhierarchical manner.

- **Pros**: High levels of redundancy and resilience; fosters a collaborative and agile environment.
- **Cons**: Complexity in setup and coordination; can lead to inefficiencies without proper governance and clarity of roles.

Hybrid Patterns

Hybrid structures combine elements from different patterns to suit specific needs (e.g., extended star, star-ring, and snowflake).

- **Pros**: Versatility and adaptability to varied requirements; potential to leverage strengths of different structural forms.

- **Cons**: Complexity in design and governance; can lead to confusion and operational challenges without clear guidelines and leadership.

The choice of structural pattern in LSS architecture is pivotal, impacting everything from operational efficiency to adaptability and resilience. Simpler structures like Bus and Star are straightforward but may lack flexibility, while more complex arrangements like Mesh and Hybrids offer robustness at the cost of increased complexity. The decision should be informed by the unique needs and goals of the LSS, balancing the trade-offs between manageability, scalability, resilience, and operational efficiency.

In the context of large sociotechnical systems, the choice of structural pattern also has profound implications for system governance. Structures like Bus and Star may centralize control, simplifying governance, but potentially creating bottlenecks or single points of failure. On the other hand, Mesh and Hybrid structures can distribute control, enhancing resilience and potentially leading to more democratic decision-making processes. Therefore, the structural pattern selection can shape not only the system's technical performance but also its social dynamics and power structures.

CHAPTER 5   ARCHITECTURE

Contextualizing the Choice

LSS architecture must not only serve the individual components but also harmonize these components toward the collective success of the entire system. At each sociotechnical level, integrative systems are vital to ensure that both the LSS as a whole and all its constituent systems optimally contribute to the overall success of the entire structure.

The Case for Extended Star

In architecting LSS, selecting an appropriate structural pattern is pivotal. The Extended Star pattern stands out as a robust and scalable model, particularly suited for organizing the diverse and interconnected systems of an LSS. This pattern features a central hub—the core integrative component—with multiple spokes extending out to various nodes, each representing a peripheral integrative component, which further connects to its respective operational component.

- **Core integrative component**: At the heart of the Extended Star structure lies the core integrative component, functioning as the central coordinating hub. Its role is to provide strategic direction, integrate diverse initiatives, and ensure overall coherence and alignment across the LSS. By centralizing strategic coordination and oversight, the core integrative component facilitates unified mission execution and seamless integration of peripheral systems.

- **Peripheral integrative components**: The peripheral integrative components act as nodes in the Extended Star framework. These components can be tailored to specific functional, regional, or divisional needs, translating the overarching strategies and directives

CHAPTER 5   ARCHITECTURE

from the core integrative component into actionable plans and operations. They play a key role in ensuring that the operational systems under their purview are in sync with the broader goals of the LSS while also allowing for a degree of autonomy and contextual adaptation.

- **Operational components**: Linked to each peripheral integrative component are the operational components. These components are responsible for the day-to-day execution of tasks and activities within their specified domain. While they operate with a certain level of independence, their actions and strategies are aligned with the directives from their respective peripheral integrative components and, ultimately, the core integrative component. This alignment ensures a cohesive approach to achieving the LSS's overall objectives.

The Extended Star pattern, with its centralized yet flexible architecture, ensures that both the core and peripheral aspects of the LSS work in harmony. The core integrative components maintain strategic oversight and integration, while the peripheral integrative components enable responsiveness and adaptability at various operational levels. Operational components, in turn, provide the necessary execution and functional capabilities. This logically hierarchical yet physically interconnected structure ensures that each segment of the LSS contributes effectively to the overall success, reminiscent of the well-coordinated and strategically aligned structure seen in efficient fleet operations.

CHAPTER 5   ARCHITECTURE

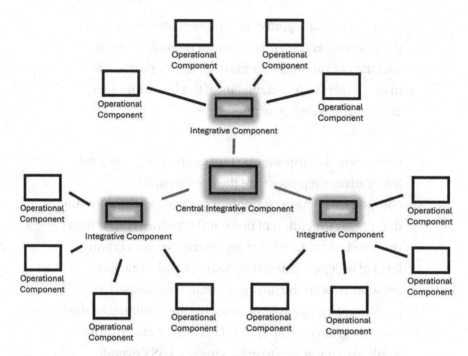

Figure 5-5. *Extended Star: balancing integration and operations in LSS*

Selecting a Prototype for LSS Architecting

In the journey to architect LSS, selecting an interesting and challenging prototype for analysis and modeling is very important because this prototype serves as a foundational model, shaping our understanding and approach to the complexities inherent in designing large-scale sophisticated systems. The choice of the right prototype is pivotal, as it influences how we perceive, interact with, and ultimately design these vast and intricate systems.

The selection of a large enterprise with a geographical divisional structure, potentially encompassing territories, regions, or districts, was made based on the following considerations:

- **Wide range of structures assessed**: In our quest for the ideal LSS prototype, we assessed various organizational structures—functional, product based, divisional, market based, geographical divisional, process based, matrix, circular, flat, and network. The geographical divisional structure emerged as the most fitting, given its complexity and relevance to modern, expansive organizations.

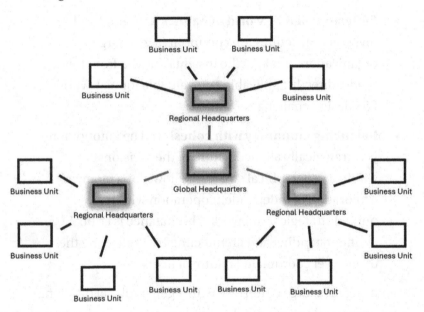

Figure 5-6. *A large enterprise with a geographical divisional structure*

- **Scalability and adaptability**: A large enterprise with a geographical divisional structure mirrors the scalability and adaptability we see in fleets. Similar to how an armada adjusts its composition by adding or removing fleets and ships, a geographical divisional enterprise can expand or contract by adjusting its regional divisions. This resemblance to the Fleet metaphor makes it an appealing choice for modeling LSS.

- **Three-level sociotechnical structure**: This structure allows for the conceptualization of an enterprise as an armada, encompassing the enterprise (armada) level, regional divisions (fleets), and individual operational units (ships). This multilevel approach is not only representative of many multinational corporations but also aligns well with the Fleet metaphor, facilitating a deeper understanding of LSS dynamics.

- **Reflection of real-world dynamics**: Geographical divisional structures are commonplace in large organizations, making the insights derived from this model directly applicable and valuable for real-world LSS design challenges.

- **Balancing autonomy with cohesion**: The autonomous yet strategically aligned nature of the divisions within a geographical structure allows for a nuanced exploration of independent operation within a unified strategic framework. This balance is critical in understanding and architecting LSS, reflecting the diverse yet coordinated nature of fleets.

By focusing on a large enterprise with a geographical divisional structure, we not only align with the Fleet metaphor but also engage with a realistic, flexible, and sophisticated model. This choice sets the stage for exploring and addressing the multifaceted challenges of architecting LSS, making it a suitable and stimulating prototype for our architectural endeavors.

Relational Patterns of LSS Architecture

Relational patterns define how systems within the LSS interact with each other and with external entities. The primary relational patterns are exchange, organizational, collaborative, cooperative, compliance, investment, and employment.

Exchange Relations

Exchange relations focus on the interactions between a system and its customers, consumers, and suppliers. These relations are transactional in nature, involving the exchange of goods, services, or resources. For customers and consumers, exchange relations often occur through self-service interfaces for accessing resources, placing orders, making payments, or providing contributions. These interactions enable users to engage directly with the system, facilitating ease of access and efficiency. For suppliers, exchange relations involve the procurement of goods or services required by the system, where terms such as pricing, delivery, and quality are negotiated and agreed upon. These interactions ensure a smooth supply chain and continuous operations by enabling efficient exchanges between the system and its suppliers.

Organizational Relations

Organizational relations are typically asymmetrical, involving directing and reporting dynamics between integrative and operational systems. In this framework, integrative systems set directions, create policies, and impose boundary constraints. In contrast, operational systems are responsible for implementing these directives and providing feedback and reports. This pattern ensures a clear hierarchy and streamlines decision-making processes within the LSS.

CHAPTER 5　ARCHITECTURE

Collaborative Relations

Collaborative relations are established to regulate system interactions in scenarios where coordinated work is necessary. These relations are characterized by joint efforts, shared responsibilities, and mutual goals. They are vital in projects or tasks requiring multiple systems to work in unison, ensuring that the activities are well coordinated and objectives are met efficiently.

Cooperative Relations

Cooperative relations are about sharing critical events and insights to enhance overall situational awareness. Unlike collaboration, which is directly related to coordinated tasks, cooperation extends to general information sharing and communication that maintain system-wide alignment and awareness. This relational pattern helps in preemptively identifying opportunities and addressing potential challenges.

Compliance Relations

Compliance relations refer to the interactions between a system and various government or special-purpose authorities like tax agencies, regulators, law enforcement, and others. These relations are governed by legal and regulatory frameworks and ensure that the system operates within the confines of established rules and standards. Compliance relations are critical for maintaining the legitimacy and legal operation of the LSS.

Investment Relations

Investment relations focus on the interactions between an enterprise and its investors, emphasizing the financial and strategic exchange that drives organizational growth. Investors provide capital to the enterprise with the expectation of financial returns, such as dividends, equity appreciation, or interest payments. These relations are built on trust, transparency, and performance, where the enterprise regularly communicates its financial health, strategic direction, and risk management through reporting and meetings. Effective investment relations ensure that the enterprise maintains investor confidence, meets legal and regulatory requirements, and continues to attract capital for future expansion and stability.

Employment Relations

Employment relations focus on the enterprise's interactions with the labor market, encompassing its ability to attract, hire, and retain talent. These relations are shaped by the enterprise's reputation as an employer, compensation structures, benefit packages, and compliance with labor regulations. Employment relations also reflect the organization's competitiveness in offering appealing opportunities for potential candidates while ensuring fair employment practices. By effectively managing employment relations, the enterprise builds its presence in the labor market, ensuring access to a skilled workforce and positioning itself as an employer of choice in a dynamic employment landscape.

Each of these relational patterns plays a vital role in defining how an LSS operates and interacts within its environment and with its constituents. By understanding and effectively implementing these patterns, LSS can achieve a harmonious balance between internal functioning and external engagement.

CHAPTER 5   ARCHITECTURE

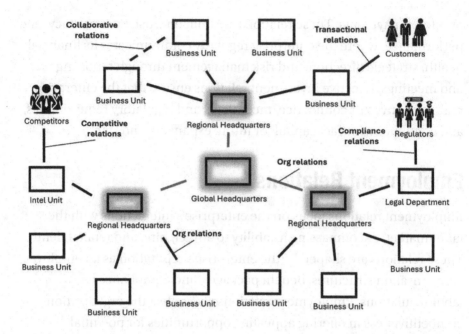

Figure 5-7. *Relationships between LSS units and between units and external entities*

Compositional Patterns of LSS Architecture

When architecting LSS, understanding and choosing the right compositional patterns is key. These patterns—Distributed, Partitioned, Interlocked, and Fused—provide varied approaches to structuring systems, each with its unique advantages and challenges.

CHAPTER 5   ARCHITECTURE

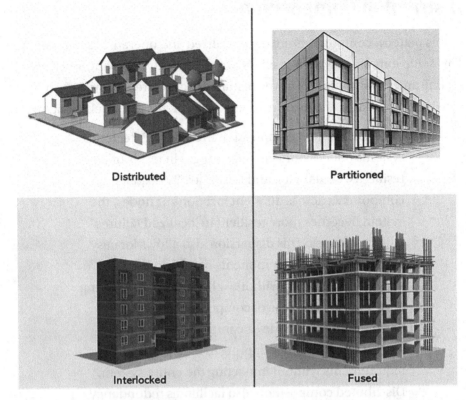

Figure 5-8. *Using a construction metaphor to illustrate four compositional patterns: Distributed, represented by detached houses symbolizing independent systems; Partitioned, shown as a townhouse with separate units; Interlocked, depicted as an apartment building where each apartment is implemented and installed as a block (module); and Fused, portrayed as a monolithic building to signify a tightly integrated system*

CHAPTER 5   ARCHITECTURE

Distributed Composition

In this pattern, components are decentralized and operate independently. This setup enhances system resilience and flexibility, allowing for easy scaling and modifications. However, managing coherence and ensuring consistent standards across distributed components can be challenging.

- **Pros**: Distributed composition in large sociotechnical systems offers significant advantages in terms of resilience, scalability, and flexibility. By dispersing components across different locations or nodes, the system becomes more resilient to localized failures or disturbances. This dispersion also allows for easy scalability, as new components or nodes can be added to the system without substantial restructuring. Furthermore, this type of composition supports flexibility, as each node or component can operate semi-independently, adapting to local conditions or requirements without impacting the entire system. Distributed composition also facilitates redundancy, where multiple nodes performing similar functions can compensate for each other in case of failure, thereby enhancing overall system reliability.

- **Cons**: Coordinating and managing a distributed system can be complex, especially when it involves numerous nodes with varying functionalities and requirements. This complexity can lead to increased overhead in communication and synchronization across different parts of the system. Moreover, maintaining consistency and ensuring timely data synchronization across distributed nodes can be challenging, potentially affecting the system's overall performance and efficiency. Additionally, as each node operates semi-independently, there might be a risk of divergent practices or standards developing, which could complicate integration and uniformity across the entire system. Security and privacy concerns are also heightened in distributed systems, as data and resources are spread across multiple locations, potentially increasing vulnerability points.

Partitioned Composition

Here, the system is divided into distinct, non-overlapping sections. This approach simplifies management by clearly delineating responsibilities and can improve system performance by isolating functions. The drawback is that it might lead to duplication of efforts and resources across different partitions, and coordinating across these partitions can sometimes be complex.

- **Pros**: Partitioned composition offers clarity and simplification by dividing a system into distinct, non-overlapping sections, each focused on specific tasks or objectives. This division not only simplifies management and understanding but also allows for specialization within each partition, optimizing efficiency and resource utilization. Moreover, the risk of system-wide failures is minimized as issues within one partition are less likely to impact others. This approach also enhances the system's modularity and flexibility, enabling easier updates and modifications within individual partitions without affecting the whole.

CHAPTER 5   ARCHITECTURE

- **Cons**: Partitioned composition presents challenges in integration and coordination across different sections, particularly when a system-wide perspective is required for changes or updates. There's a potential risk of resource duplication across partitions, leading to inefficiencies at the larger system level. Additionally, ensuring effective communication and data exchange between partitions can be difficult, potentially leading to isolated information silos. This composition may also limit the system's ability to adapt quickly to holistic changes, as each partition tends to focus on its immediate scope, sometimes at the expense of a broader, more integrated perspective.

Interlocked Composition

This pattern features components that are closely linked, with each part's operation directly affecting the others. It allows for efficient resource sharing and tight coordination, leading to optimized performance. The interdependency, however, might introduce complexities in troubleshooting and can lead to systemic vulnerabilities—a failure in one component could impact the entire system.

- **Pros**: Interlocked composition in LSS presents several key advantages, particularly in terms of integration and efficiency. This approach, where different components or modules are tightly interconnected, ensures a high degree of coordination and interaction among various parts of the system. Such integration facilitates the seamless flow of information and resources, leading to increased operational efficiency and synergies. Interlocked composition also promotes consistency

CHAPTER 5 ARCHITECTURE

and uniformity in processes and practices, as the tightly coupled nature of the system components necessitates standardized procedures and interfaces. This composition type is beneficial in contexts where close collaboration and immediate response among different system elements are required.

- **Cons**: On the downside, interlocked composition can bring about certain limitations and challenges. The tight coupling between components can lead to rigidity, making it difficult to modify or scale individual parts of the system without affecting the whole. This can limit the system's adaptability to changing requirements or environments. Additionally, the interdependence of components means that a failure in one part can have a cascading effect on the rest of the system, potentially leading to widespread disruptions. The complexity of managing and maintaining such an intricately connected system can also be higher, requiring sophisticated coordination mechanisms and oversight. Furthermore, this type of composition may stifle innovation in individual modules or components, as the focus on maintaining seamless integration can limit the scope for independent development or experimentation.

Fused Composition

Fused composition indicates a high level of integration where boundaries between components are blurred. This pattern facilitates seamless operation and unified management of resources but can make adaptation

CHAPTER 5   ARCHITECTURE

and modular changes more difficult. It also requires careful initial planning and design to avoid entangled dependencies that are hard to manage or modify.

- **Pros**: Fused composition offers distinct advantages, particularly in terms of cohesive functionality and robustness. In this composition style, different components are not just interconnected but essentially merged into a unified whole. This results in a highly integrated system where the boundaries between components are blurred, leading to a seamless user experience and operational efficiency. Fused composition ensures that all parts of the system work in harmony, with minimal friction or mismatch. This integration often leads to optimized performance, as the system is fine-tuned to operate as a single entity. It's especially effective in environments where a streamlined, unified approach is paramount for success.

- **Cons**: Fused composition also has its drawbacks. The main challenge lies in its inflexibility; once components are fused, modifying any single element becomes a complex task that often requires rethinking or adjusting the entire system. This lack of modularity can hinder adaptability and scalability, as changes to one part of the system can necessitate extensive reengineering of the whole. Moreover, the tight integration of components can make troubleshooting and maintenance more challenging, as identifying and isolating issues within a fused architecture is often more complex than in more modular setups. The risk of overdependence on single points of failure increases, as the entire system can be more vulnerable

CHAPTER 5   ARCHITECTURE

if key integrated components encounter problems. Additionally, the creativity and innovation potential within individual components might be limited, as the focus shifts to maintaining the integrity and functionality of the combined structure.

Compositional Pattern Selection in LSS

The selection of compositional patterns within large sociotechnical systems architecture is closely tied to the levels and goals of the system. At the highest level, corresponding to the Fleet, a Distributed composition is most suitable, facilitating broader oversight and easier scalability. Moving to the Ship level, a Partitioned approach becomes more appropriate, allowing for defined yet autonomous functional areas within a single entity. At the more granular levels, such as within a Ship's block or module, Interlocked and Fused compositions are advantageous. These styles offer tighter integration and cohesive functionality, essential for the detailed operations and interactions within these smaller, more intricately connected segments. This tiered application of compositional styles, each aligning with its respective level's requirements and objectives, ensures a well-balanced and effective architectural structure for the LSS.

Relating LSS to Maritime Formations

The metaphor of a fleet, particularly an armada—the largest maritime formation—provides a powerful lens through which to view and design LSS for a global enterprise with a geographical divisional structure. Drawing parallels between a modern global enterprise and an armada offers profound insights. An armada, a grand naval formation comprising multiple fleets, mirrors the structure and dynamics of a multinational

CHAPTER 5   ARCHITECTURE

corporation. This metaphor not only enriches our understanding of LSS architecture but also provides a tangible, comprehensive framework for conceptualization and design.

To grasp the depth of this metaphor, let's explore how its various facets can be leveraged in architecting an LSS that is resilient, strategically aligned, and operationally agile.

Armada As a Metaphor for Global Enterprises

In the armada metaphor, the global enterprise is envisioned as a fleet of fleets—an expansive, multifaceted force wherein the central enterprise integrative unit (like the command ship) leads geographical divisions (analogous to regional fleets). This framework not only visualizes the magnitude but also the composite and multifaceted nature of multinational corporations.

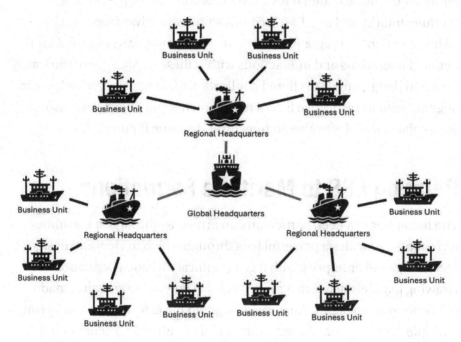

Figure 5-9. *A global enterprise as an armada*

CHAPTER 5   ARCHITECTURE

The flagship, within the armada metaphor, represents the corporate headquarters, but it's essential to recognize that in the vast maritime landscape, it's just one ship among many. It symbolizes a distinctive social structure, led by the CEO and expanded through technological means—the ship itself. This viewpoint is key: the command ship, while vital, is not a solitary entity reigning from afar. Instead, it's similar to an admiral's vessel. The CEO, much like an admiral, is not a distant figure swimming alone in the ocean of corporate strategy, indulging in substantial salaries and bonuses in isolation. They are an integral part of a larger crew, guiding the ship together with their team. Their leadership is deeply entrenched within the ship's operations, highlighting the collaborative nature of navigation and strategy. This approach emphasizes the importance of joint effort and unity in directing the corporate armada, underscoring that every vessel and crew member, from the executive suite to the operational staff, plays a critical role in navigating the corporate seas.

CHAPTER 5 ARCHITECTURE

Figure 5-10. *All hands on deck: Evolveburgh's CEO proves that "flagship" means navigating alongside Sales, HR, and Tech—because no one's strategizing from a yacht*

Regional Fleets Within the Armada

Each regional fleet, while autonomous in operational tactics, aligns with the armada's broader strategic goals. This reflects the balance between localized decision-making and adherence to central corporate strategy in multinational firms.

CHAPTER 5   ARCHITECTURE

Mirroring the diverse capabilities and specializations of ships within a fleet, geographical divisions are tailored to specific market needs and local business environments, demonstrating the necessity of contextual adaptability within a unified strategic framework.

Within the context of the armada metaphor, each regional fleet, representing a geographical division of the global enterprise, is guided by its own command ship, similar to regional integrative units. These command ships play an important role in the broader structure of the enterprise armada acting as the integrative unit for its respective division. They are tasked with translating the overarching strategic directives from the flagship (the corporate headquarters) into actionable plans and operations that resonate with the local markets and regulatory, cultural, and other landscapes. In doing so, they maintain a delicate balance between aligning with global strategies and exercising the autonomy needed for regional effectiveness and responsiveness.

The flagship, along with these regional command ships, forms *the integrative core* of the enterprise. This central integrative core is instrumental in directing the operations of the peripheral units—the ships or operational teams within each regional division. The regional command ships, therefore, are not only executors of the central strategy but also key contributors to the strategic decision-making process, offering insights and feedback from their unique regional perspectives. This setup ensures that strategies are not just top-down impositions but are informed by on-ground realities across different markets.

CHAPTER 5   ARCHITECTURE

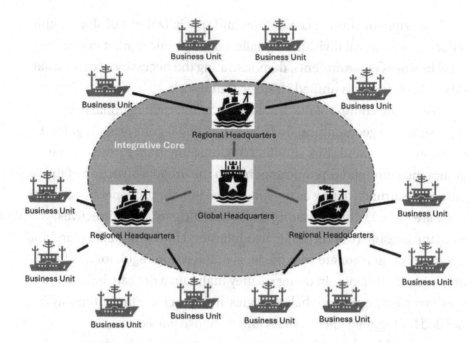

Figure 5-11. The flagship, along with the regional command ships, forms the integrative core of the enterprise

This interplay between the flagship and the regional command ships underscores a collaborative dynamic within the enterprise's leadership. It mirrors the cooperative nature of an armada, where the flagship and individual fleet commanders work in concert to navigate complex, often unpredictable environments. This cooperative framework allows for a more agile, responsive, and context-aware approach to managing global operations, ensuring that the entire armada moves cohesively toward shared objectives while valuing the unique contributions of each division.

In essence, the regional fleets and their command ships are pivotal in ensuring that the global strategy is effectively localized and implemented, while their constant interaction with the flagship guarantees that the enterprise's overall direction remains cohesive and strategically sound.

CHAPTER 5　ARCHITECTURE

This arrangement facilitates a robust, adaptable, and resilient enterprise structure, capable of thriving in the dynamic waters of international business.

Ships Within the Armada

Moving deeper into our maritime metaphor, we arrive at the individual ships that form the essence of both fleets and the armada at large. While we conceptualize an enterprise as an armada, and each regional division as a fleet, it's critical to recognize that, fundamentally, the armada is a composition of ships—both integrative and operational. This perspective brings several strategic advantages in architecting and managing large sociotechnical systems.

- **The network of ships—a flat formation**: Despite the apparent hierarchical imagery an armada might conjure, at its core the structure is a network—a flat formation of ships. Each ship, whether it serves an integrative or operational role, contributes to the armada's overall mission. This networked structure simplifies how the organization is modeled, communicated, and understood. By viewing the enterprise as a network of ships, we break down complex, multilayered hierarchies into more manageable, interconnected units. This flat formation aids in reducing bureaucratic layers, fostering quicker decision-making, and enhancing agility. It allows for easier adaptation to changing conditions, mirroring the way a fleet of ships can quickly rearrange itself to meet emerging challenges or opportunities.

CHAPTER 5   ARCHITECTURE

- **Standardization and modularization in ship design:** Drawing from contemporary naval architecture, the design of each ship (enterprise unit) in the armada can be inspired by the principles of modularization and containerization. Just as modern ships are built with standardized parts and modular containers for efficient, scalable, and flexible cargo management, enterprise units can adopt a similar approach. Standardized designs across different units (ships) promote uniformity, making it easier to manage, operate, and maintain them. Modular elements within each unit allow for customization and flexibility, adapting to specific tasks or regional needs without disrupting the overall system. This approach facilitates scalability—just as a ship can carry different types of cargo containers, an enterprise unit can adjust its functions and processes based on immediate needs and strategic shifts.

- **Enhanced scalability and flexibility:** The containerization approach in shipbuilding, where containers of standard sizes are used to transport goods globally, enabling efficiency and flexibility, can be metaphorically applied to managing business functions and processes. In enterprise terms, this means creating standardized, container-like modules (processes, practices, teams) that can be easily moved, replaced, or modified without affecting the integrity or the operational capacity of the whole. This modular design ensures that each unit of the enterprise can rapidly adapt to new technologies, market demands, or organizational changes, much like how a cargo ship can be swiftly reconfigured to carry different types of cargo.

CHAPTER 5   ARCHITECTURE

- **Improved comprehension and communication**:
  A system modeled on the clarity and uniformity of a fleet of ships allows for straightforward understanding and communication of the organizational structure and functions. When each unit is seen as a ship with a specific role and capability, it becomes easier for individuals within the organization to perceive how different parts fit into the larger picture, how they interact, and how their contributions drive the collective mission forward.

Figure 5-12. Breaking tradition: Evolveburgh's council mandates a "flat armada" structure—because nothing says unity like a perfectly synchronized fleet pretending hierarchy doesn't exist

207

In summary, by envisioning each unit of the enterprise as a ship within a flat, networked armada, we achieve a model that is not only easy to comprehend and communicate but also robust in its standardization and flexibility. This approach mirrors the advanced design of modern fleets, ensuring that the enterprise remains resilient, adaptable, and poised for efficient operation in the dynamic seas of global business.

Architectural Implications

Adopting the fleet metaphor for architecting LSS naturally guides the selection of patterns for component selection, structural organization, strategic positioning, and logical interrelations.

- **Selection pattern—Homogeneous**: Drawing from the fleet metaphor, adopting a homogeneous selection for system components across the LSS fosters uniformity and predictability. This similarity simplifies management and integration, similar to a fleet where ships share common designs and capabilities to enhance interoperability.

- **Structural pattern—Extended Star topology**: An Extended Star topology, similar to a flagship and its fleet, is optimal for balancing centralized command with operational distribution. This structure facilitates clear decision-making and efficient information flow, embodying the strategic and operational dynamics of an armada.

- **Relational pattern(s)—Spectrum of relationships**: The metaphor supports various relational patterns, from transactional to cooperative interactions. This versatility mirrors the fleet's capability to adapt its relational dynamics based on operational context, ensuring the LSS is flexible and responsive to different scenarios.

- **Compositional pattern(s): Distributed composition** for scalability and oversight at armada/fleet level. **Partitioned composition** for functional autonomy within integrative and operational units, reflecting how different sections of a ship operate independently yet cohesively at the ship level. **Interlocked** and **Fused compositions** for detailed integration and functionality necessary for complex operations within smaller segments at the Block/Module Level.

Modeling a global enterprise as an armada offers a structured yet flexible metaphor for comprehending and crafting LSS. This analogy provides rich insights for strategic command and operational agility, equipping architects with a robust framework to develop sociotechnical architectures that are not only resilient and adaptable but also clear and intuitive. The fleet metaphor, with its layered and interdependent structure, is invaluable for envisioning and implementing the complex architecture of modern digital enterprises.

Novel Architecture Style: Unit Oriented Enterprise Architecture

In the evolving landscape of large sociotechnical systems, there is an emerging need for architectural styles that not only address complex and dynamic business requirements but also encapsulate the agility and resilience of modern enterprises. It is in this context that we introduce a novel architectural style: Unit Oriented Enterprise Architecture. This style, inspired by the robust and flexible structure of maritime formations, particularly the structural elegance and organizational dynamics of a fleet, aims to deliver a harmonious alignment of business objectives with technological capabilities.

CHAPTER 5   ARCHITECTURE

Embodying Business-Technology Alignment

UOEA introduces an approach where the harmony between organizational units and information technology is paramount. This harmony is achieved by aligning the social architecture of the organization—its people, culture, and operational workflows—with the digital architecture, comprising the various technological tools and platforms. In UOEA, each social unit of the enterprise is extended and enhanced by corresponding digital technologies. This extension is not just a mere addition of technology to existing social structures, but a thoughtful integration where social and digital domains enrich and amplify each other.

This alignment ensures that technological solutions are not just layered on top of business processes, but are seamlessly integrated into the very core of the fleet's organizational structure, making crews and technology a cohesive unit. By doing so, technology becomes a natural extension of the social units, facilitating more intuitive and effective workflows, decision-making, and problem-solving processes. This integration also means that changes in business strategy or operations are immediately mirrored in the digital infrastructure, allowing for a cohesive and agile response to new challenges and opportunities. Thus, UOEA serves not just as an architectural framework but as a strategic approach to aligning and harmonizing the dual realms of business and technology, social and digital, for optimum effectiveness and growth.

Ensuring Clear Ownership and Control

UOEA emphasizes clear ownership and control of each unit. Just as every ship within a fleet has a captain responsible for its operations, every unit in UOEA is designed with clear governance and accountability. This clarity in ownership ensures that decision-making is swift and responsible, attributes essential for the agility and effectiveness of LSS.

Enhancing Clarity and Comprehensibility

The modern digital enterprise often presents a labyrinthine complexity in its digital structures, creating significant challenges in terms of visibility and comprehensibility for architects, designers, and stakeholders. In the dense weave of intertwined systems, processes, and data flows, the clarity of these structures can become obscured, leading to inefficiencies and increased potential for misalignment between technology and business objectives.

Drawing inspiration from the elegance and structured simplicity of fleets, UOEA offers a powerful antidote to this prevailing obscurity. In naval terms, each ship in a fleet possesses a distinct role, clear functionality, and a well-understood place within the larger formation. By modeling digital enterprise structures on this maritime paradigm, UOEA facilitates a clearer and more comprehensible architecture.

This metaphorical alignment with fleets allows for an intuitive visualization and understanding of the digital enterprise. Just as one can readily discern the function and importance of different ships within a fleet, so can architects and stakeholders perceive and comprehend the various digital components of an enterprise. Each unit—analogous to a ship—is distinct yet integrated, with its purpose, capabilities, and relationship to the whole readily apparent.

This approach demystifies the digital architecture of large enterprises. It breaks down complex, often opaque digital structures into manageable, clearly defined units. Architects and designers, therefore, gain the ability to effectively map, analyze, and optimize these structures. Stakeholders, on their part, can more easily grasp the operational dynamics and strategic positioning of different technological elements within the enterprise. In doing so, UOEA not only enhances the visibility of the digital structure but also fortifies the alignment between technology and business, driving more effective decision-making and strategic planning.

CHAPTER 5   ARCHITECTURE

Facilitating Changeability: Scalability, Transformability, and Flexibility

In the fast-paced and often unpredictable global business environment, the ability to scale, transform, and flex according to the need is indispensable. UOEA, drawing inspiration from the modular nature of fleets, allows for such scalability and flexibility. Units can be rapidly reconfigured or scaled out and in, aligning seamlessly with fluctuating demands and evolving enterprise environments.

Streamlining Communication Flows

The structural layout of UOEA ensures that communication within the LSS is streamlined and effective, similar to the communication channels within a well-orchestrated fleet. This streamlined communication aids in quick information dissemination and decision-making, key elements in maintaining the operational efficiency and responsiveness of the system.

Unit Oriented Enterprise Architecture is more than just a theoretical construct; it's a pragmatic approach tailored for modern enterprises facing multifaceted challenges and opportunities. By adopting the principles of UOEA, large sociotechnical systems can achieve a dynamic balance between robustness and agility, control and innovation, centralization and decentralization—essentially encapsulating the paradoxical demands of the contemporary business environment. This architecture style, therefore, marks a significant step forward in the way we conceptualize, design, and implement large-scale systems in an increasingly complex world.

Unit Oriented Enterprise Architecture: Key Definitions

System

Definition: A system is a regularly interacting or interdependent group of elements forming a unified whole.

Explanation: This definition emphasizes not only the individuality and distinct functionality of each element but also the integral whole-part relationship within the system. Each element operates independently but is inherently connected to and shaped by the larger system's objectives and strategies. The dynamic interplay between each element and the overall system ensures a cohesive, integrated functionality, reflecting the interdependence and harmonious functioning of components in a robust, responsive structure.

Social System (or Collective)

Definition: A social system is an interrelated group of individuals forming a unified whole and regularly interacting within a bounded environment for fulfilling a common purpose.

Explanation: This term refers to a group of people who are connected through social relationships, interacting within specific boundaries, such as an organization, organizational unit, or community.

Their interactions are usually guided by a shared objective or goal that requires contributions of all members.

Sociotechnical System

Definition: A sociotechnical system is a social system with physical or digital extensions.

Explanation: This expands upon the concept of a social system by incorporating technological components. These systems acknowledge the interplay and mutual influence between social groups (humans) and their technology (tools, software, machines), recognizing that these elements co-evolve and deeply affect each other's functionality and evolution.

Large Sociotechnical System (LSS)

Definition: A large sociotechnical system is a system of sociotechnical systems.

Explanation: This refers to a more complex and extensive configuration where multiple sociotechnical systems are interconnected and interact with each other, forming a larger, more complex network. Examples include large corporations or urban infrastructures that comprise various interlinked sociotechnical systems.

Crew

Definition: A crew is the social component of a sociotechnical system.

Explanation: Here, the focus is on the human aspect within the sociotechnical system. The crew signifies the group of individuals who operate, manage, or interact with the other components (technological or otherwise) of the system.

Craft

Definition: The craft is the main digital extension of a sociotechnical system. Other digital extensions of the sociotechnical system are smart (Internet of Things) objects.

Explanation: In the context of UOEA, the craft is similar to a ship, aircraft, or spacecraft, serving as the primary technological structure that facilitates the functioning of the crew. It acts as an integrative framework, combining all the technological tools and systems that support human activities, with the exception of smart objects. This integration ensures that the technological aspect of the sociotechnical system is cohesive, efficient, and fully supportive of the crew's needs, much like how the physical structure of a craft seamlessly incorporates various technologies for its operation and mission success.

CHAPTER 5   ARCHITECTURE

Unit

Definition: A unit is a sociotechnical system, a collective extended by digital technology. Unit = Crew + Craft + Other Extensions.

Explanation: In UOEA, the unit is a fundamental concept central to the entire architectural framework. UOEA emphasizes the design of LSS at two distinct but interconnected levels. At the topmost level, an LSS is conceptualized as a flat network comprising multiple units. This approach contributes to the overall elegance and comprehensibility of the architectural style, as it simplifies the visualization and management of complex systems into more manageable, interconnected units. At the subsequent level, attention is focused on the internal architecture of each unit. Here, UOEA stresses the importance of harmoniously aligning the social system (crew) with the technological structure (craft), ensuring that each unit is both functionally effective and architecturally coherent. This dual-level focus facilitates a clear, streamlined design approach, where the macro-architecture of the LSS and the micro-architecture of individual units are both integral and complementary components of the overall system architecture.

Agent

Definition: An agent is an individual extended by digital technology. The capabilities of human agents can be augmented by artificial agents, virtual assistants, tools, and smart objects.

Explanation: This concept captures the notion of human individuals augmented or assisted by various forms of technology. These augmentations can range from simple tools to sophisticated artificial intelligence, enhancing the individual's abilities to perform tasks, make decisions, or interact with other components of the system.

Unit Oriented Enterprise Architecture: Core Principles

In the pursuit of architecting robust and resilient LSS, UOEA distinguishes itself through a set of foundational principles that serve as the bedrock for designing systems that are not only efficient and scalable but also aligned with the complex interplay of social dynamics and digital innovation.

Together, these principles guide the creation of an architecture that is not only aligned with the enterprise's strategic vision but also flexible and adaptable to the inevitable shifts in the technological landscape and organizational needs.

CHAPTER 5   ARCHITECTURE

Principle 1: Harmonious Integration of Realms

What: LSS must harmoniously integrate social and digital realms.

Why: This principle ensures a seamless interplay between human and technological elements of an LSS. It recognizes that the strength of a sociotechnical system lies in the symbiotic relationship between its social structures and digital technologies, enhancing the overall performance and adaptability of the system.

Principle 2: Digital Within Social Legitimacy

What: Digital components are nested within the social component's boundary.

Why: Positioning digital technology within the boundary of social legitimacy ensures that technological developments and implementations are accountable to, and in harmony with, the social system's needs and norms, fostering responsible and effective use of digital structures.

Principle 3: Sociotechnical Focus on Social and Digital Realms

What: UOEA prioritizes the social and digital realms, leaving physical realm considerations to traditional architecture.

Why: Concentrating on the dynamic and rapidly evolving social and digital realms ensures the architecture remains flexible and responsive to change, while the more static physical realm is managed by established architectural practices.

Principle 4: Purpose-Driven Sociotechnical Wholeness

What: Each system within an LSS represents a unified whole aligned with a specific purpose.

Why: By focusing on purpose driven design, this principle ensures that every component within the system contributes to a coherent purpose, facilitating a clear and directed orchestration of both social and digital activities. Ultimately, purpose driven design provides a clear path for each unit to contribute meaningfully to the LSS's overarching objectives, ensuring that the system's complexity does not hinder but rather enhances its capability to adapt and thrive in dynamic environments.

Principle 5: Coherence in Purpose Implementation

What: Each sociotechnical unit should exhibit a coherent implementation chain that includes function, process, structure, order, culture, memory, and storage, all geared toward fulfilling the unit's purpose.

CHAPTER 5 ARCHITECTURE

Why: This principle highlights the importance of a well-synchronized operational system where every digital and social component is methodically aligned to serve the unit's purpose. Such coherence ensures that the functional interfaces and activities, procedural flows, structural arrangements, rules and constraints, cultural dynamics, history, and digital storage collectively translate the unit's overarching goals into tangible outcomes. It is this purpose-driven coherence that transforms a mere collection of digital structures into a synergistic and functional whole, reinforcing the unit's role within the larger sociotechnical system. The term "implementation" in the title encapsulates the active, dynamic process of turning strategy into action, where the focus is on the actualization of objectives through a unified and supportive framework of system components.

Principle 6: Flat Network Organization with System-of-Systems Relationships

What: An LSS may have nested system-of-systems relationships, but it must operate as a flat network of sociotechnical units organized using an extended star topology.

Why: While hierarchical relationships may exist conceptually, a flat network structure ensures agility and reduces the complexity of interactions. The extended star topology allows for centralized coordination, when necessary, while preserving

CHAPTER 5   ARCHITECTURE

the autonomy of individual units. This organization mimics the responsiveness and robustness found in decentralized natural systems.

Architecting the Enterprise: Achieving Global Reach

In an era where business dynamics are globally interconnected, architecting an enterprise for global reach is not just advantageous—it is imperative. UOEA provides a comprehensive framework for enterprises to not only expand their operations across diverse geographies but also to resonate with local markets and regulatory environments effectively. Let's explore the transformative strategies within UOEA that enable an enterprise to harmonize its overarching goals with the nuances of global stakeholders. From deploying stakeholder-specific units that address local demands to weaving a cohesive global strategy that aligns with the enterprise's core mission, this section illuminates the path toward a robust global presence. Through UOEA's lens, we will dissect the essential components that constitute a globally oriented LSS, offering insights into purpose driven design, compositional flexibility, structural topology, and the nuances of inter-unit relationships—all contributing to a sustainable and scalable global enterprise architecture.

Global Reach

The concept of "Global Reach" in the UOEA framework is fundamentally about understanding and structuring the enterprise to respond to a range of global stakeholder demands. By partitioning the global environment

CHAPTER 5   ARCHITECTURE

into entities that represent diverse stakeholder groups, UOEA provides a lens through which the enterprise can view its operations from a multi-stakeholder perspective. These entities include but are not limited to

- **Job markets**: These represent the employees and potential employees influencing the enterprise's human resource strategies.
- **Consumer markets**: These encapsulate the customers, shaping the enterprise's product and service offerings.
- **Financial markets**: These involve shareholders and creditors, guiding the enterprise's financial strategies and reporting.
- **Supplier markets**: These represent the suppliers, impacting the supply chain and procurement strategies.
- **Public and government agencies**: These include tax authorities and regulators, dictating compliance and public engagement strategies.

When an enterprise is cognizant of these entities, the architecture inherently becomes stakeholder focused. This allows the enterprise to craft units with clear, stakeholder-oriented purposes. For instance, a unit dedicated to the job market may focus on talent acquisition and retention, aligning its strategies with global employment trends and regional labor laws.

Defining these clear purposes enables each unit within the enterprise to align its operations with the particular demands and expectations of its respective stakeholder group. A consumer market–oriented unit would prioritize customer satisfaction and product innovation, while a financial market–oriented unit would focus on fiscal responsibility and investor relations.

CHAPTER 5   ARCHITECTURE

This strategic partitioning affects the LSS structure by establishing a framework for global reach that is responsive and adaptive. Each unit, serving as an emissary to its stakeholder entity, can act independently to meet localized needs while collectively contributing to the enterprise's overarching global strategy. For example, a unit in one region may adapt its marketing strategy to suit local consumer preferences while still adhering to the global brand strategy.

In practice, this approach transforms how an enterprise extends its reach. Rather than a monolithic entity trying to blanket disparate markets with a one-size-fits-all strategy, the enterprise becomes a constellation of specialized units. Each is fine-tuned to operate effectively within its designated stakeholder environment, ensuring that every unit contributes both to regional success and the global whole.

By enabling an enterprise to resonate with stakeholders' local nuances while maintaining a unified global presence, UOEA's stakeholder-oriented partitioning becomes a powerful tool for achieving comprehensive global reach.

Selecting LSS Components

In UOEA, the selection of components that form the LSS is critical for establishing a structure that is robust, responsive, and aligned with strategic objectives. Let's delve into the criteria and characteristics of these components, which are central to achieving operational excellence and strategic coherence across the enterprise:

- **Sociotechnical components as exclusive LSS building blocks**: At the heart of the LSS are sociotechnical components, or *units*, each embodying the principles of legitimacy, accountability, and purposefulness. These units are the essential building blocks of the system, and it is through their strategic design and dynamic interrelations that they coalesce ensuring the entirety of the LSS embodies these core principles. The legitimacy of each unit is derived from its alignment with the organization's ethos and regulatory adherence, accountability stems from clear lines of responsibility and transparent communication, and purposefulness is ingrained in the unit's mission to fulfill specific stakeholder needs.

- **Differentiating integrative and operational units**: LSS units are of two main types—integrative and operational. Operational units are the engines that drive the organization's primary activities, directly engaging with and serving stakeholders such as customers, employees, or suppliers. On the other hand, integrative units are the cohesive force that unites these operational components, ensuring the sum of the parts functions as a coherent whole. For a geographically dispersed enterprise, the global integrative unit acts as the flagship, guiding the collective direction, while regional integrative units—much like command ships within a fleet—orchestrate the efforts of operational units within their purview, tailoring strategies to regional dynamics and needs.

- **Ensuring uniform structure across units**: To maintain coherence and efficiency, each unit within the LSS is uniformly structured. This uniformity translates into a consistent framework of operation and interaction, enabling seamless integration and coordination across the enterprise. Every unit is a symbiotic amalgamation of a social crew and digital technology; the crew has full control over their digital craft and remote digital objects, fostering a sense of ownership and responsibility. This control is not merely operational but extends to the very architecture of the digital tools at their disposal, granting them the ability to adapt and evolve their tools as needs and circumstances dictate.

In this way, the careful selection and characterization of LSS components under UOEA principles not only bolster the individual effectiveness of each unit but also contribute to the systemic resilience and agility of the enterprise as a whole. This strategic component selection ensures that each unit, whether operational or integrative, is fully equipped to fulfill its role within the larger mission of the enterprise, thereby facilitating a robust global reach while maintaining local relevance and impact.

Composing the Large Sociotechnical System

- **Distributed architecture and loose coupling**: In the design of LSS, a distributed architecture is not just a choice but a strategic imperative. This architecture mirrors the organizational structure of a maritime fleet, where extreme loose coupling prevails. Each ship within a fleet operates as an independent entity, capable of navigation and mission fulfillment, yet remains part of a larger coordinated group. This approach brings substantial benefits, including heightened agility, where each sociotechnical unit within the LSS can swiftly respond to changes without the need for systemic overhaul. It also increases fault tolerance, as issues within one unit do not cascade throughout the system.

- **Decentralization and autonomous operation:** Decentralization is a hallmark of this approach, granting autonomy to each unit similar to ships in a fleet. Each unit, or "ship," is self-contained, with its own crew and digital constructs and resources, allowing for operations to continue independently of central command under most circumstances. This autonomy empowers local decision-making, which is often more responsive to immediate situational demands than centralized governance structures could be.

With distributed architecture, the enterprise's capability to integrate or disintegrate units or formations as needed is not merely a strategic advantage but a testament to the resilience and adaptability ingrained in the LSS's very architecture. This flexibility is the cornerstone of maintaining a robust global presence and ensures that the enterprise, much like a well-commanded fleet, can navigate through any storm and emerge ready for the voyage ahead.

- **Mergers and acquisitions:** In the context of mergers and acquisitions, the modular nature of a distributed LSS allows an enterprise to seamlessly integrate new formations of units—or disintegrate them if necessary—much like an armada assimilating new squadrons. This integration extends beyond simply adding assets; it involves aligning operational models, cultural elements, and strategic objectives. Each unit or new "squadron" is carefully interwoven into the existing structure, ensuring that it complements and enhances the collective capabilities of the enterprise without causing disruption to established workflows.

- **Market shifts and competitive moves**: When market dynamics shift or when an enterprise faces new competitive challenges, individual units can act independently to recalibrate their strategies. This decentralized agility enables them to respond proactively to external pressures, adapt their operational focus, or innovate their service offerings. In doing so, they can maintain market relevance and competitive advantage without the need for cumbersome system-wide reorientations.

- **Technology changes**: Technology evolves at a relentless pace, and an enterprise's ability to stay current is critical. Within a distributed LSS, each unit has the autonomy to adopt new technologies, phasing out obsolete systems, and implementing cutting-edge solutions. This unit-level technological adaptability ensures the enterprise remains at the forefront of innovation, with each "vessel" in the fleet upgrading its navigational tools to traverse the digital seas effectively.

- **Unique customer requirements**: The flexibility inherent in UOEA allows individual units within an enterprise to specialize and adapt their outputs to meet unique customer requirements—adjust their offerings, optimize their engagement strategies, and pursue localized initiatives. When customers present needs that diverge from the norm, the relevant units can modify their operations, leveraging their autonomy to create customized solutions. This ability to pivot and cater to specific needs demonstrates the system's inherent responsiveness, ensuring customer satisfaction and loyalty without disrupting the broader operational efficiency of the enterprise.

- **New regional opportunities**: When an enterprise identifies an opportunity to enter a new regional market, the UOEA framework facilitates the establishment of a new regional formation. This is similar to deploying a new fleet to uncharted waters. The enterprise can construct a tailored presence in this region, with units specifically designed to address the unique cultural, legal, and economic contexts of the market. As this regional formation takes shape, it operates with a degree of independence, yet remains aligned with the global strategy of the enterprise. The capability to add new regional formations enables an enterprise to scale and expand its global reach methodically aligning local presence with global objectives.

- **Regulatory updates**: Regulatory changes can necessitate rapid adjustments within an enterprise. In a distributed LSS, the units responsible for compliance can implement regulatory updates and disseminate the necessary procedures across the relevant units swiftly and efficiently. This approach reduces the regulatory burden and ensures compliance without disrupting the enterprise's broader operational rhythm.

- **Internal reorganizations**: Finally, internal reorganizations within an LSS can be approached with precision and minimal disturbance. Changes can be confined to specific units or formations that require restructuring, allowing the rest of the enterprise to continue its operations undisturbed. This localized approach to reorganization minimizes downtime and maintains overall system integrity, ensuring the enterprise remains dynamic and cohesive throughout the transition.

CHAPTER 5   ARCHITECTURE

Structuring the LSS

The structuring of a large sociotechnical system is pivotal for its functionality, flexibility, and efficiency. Adopting the appropriate topology for an enterprise is similar to determining the formation of a naval force, be it a fleet or an armada, with careful consideration for the size and complexity of the organization.

Basic Star Topology for Midsize Enterprises: The Enterprise As a Fleet

For midsize enterprises, the basic Star topology serves as an effective framework. In this configuration, the enterprise is viewed as a fleet, where there is a central integrative unit—similar to the flagship—that provides the strategy and direction. This flagship is directly connected to a range of operational units—ships—each of which is responsible for executing a specific aspect of the enterprise's operations. This structure enables a clear line of communication and control from the center, facilitating coordination and quick response to the central unit's directives while maintaining the operational autonomy of the peripheral units.

Enterprise As an Armada: Formation of Fleets (Nesting)

When the enterprise scales up to a more complex global operation, it transitions into what can be described as an armada—a formation of fleets. Each fleet, led by its fleet integrative unit (command ship), operates within a particular region or domain, coordinating a group of operational units (ships). This nesting allows for a nuanced approach to governance and strategy deployment, where regional nuances are managed by fleet integrative units, yet still align with the global vision set by the armada integrative unit (flagship).

CHAPTER 5   ARCHITECTURE

Enterprise As an Armada: Formation of Fleets and Flotillas (Double Nesting)

In the most intricate enterprises, a double-nesting extended Star topology may be employed. Here, the enterprise-armada comprises multiple fleets, and within each fleet are several flotillas, each a formation of operational ships. The flotilla integrative units manage smaller, more focused areas of operation, ensuring an even closer alignment with local market needs and operational challenges. This arrangement also supports specialization within a diverse global enterprise structure.

The Outcome: A Flat Network of Ships

Despite the complexity introduced by nesting, the organizational aim is to maintain an actual flat network of ships. This ensures a streamlined hierarchy where, to an external observer, the intricacies of flotillas may be imperceptible, reflecting a structure that is simple to navigate and manage. This architectural approach simplifies the comprehension of the enterprise's structure, regardless of whether it is organized as a fleet, a formation of fleets, or a formation of fleets and flotillas. The integrative core (armada, fleet, and flotilla integrative units) represents the central control mechanism that is vital for maintaining the overarching strategy and governance. At the same time, the operational periphery (operational ships) ensures that the enterprise remains agile, innovative, and closely connected to the local demands and stakeholder needs.

Balancing Central Control and Local Autonomy

This balance between central control and local autonomy is vital for the global reach of the enterprise. It enables the enterprise to act cohesively toward its strategic goals while remaining flexible enough to adapt to local market demands and dynamic global trends. The integrative core ensures a unified direction and purpose, while the operational periphery allows for differentiation in execution, reflecting the diversity of the environments in which the enterprise operates.

Interacting with Entities in the Surrounding Environment

The structure of the LSS facilitates effective interactions with various entities in the environment. By clearly delineating the roles and scopes of operation for the core and periphery, the enterprise can engage efficiently with job markets, consumer markets, financial markets, supplier markets, and public and government agencies. The operational units can interact directly with these entities, drawing from the strategy and support of their integrative units to ensure that all engagements are purposeful and aligned with the enterprise's global objectives.

Whether an enterprise resembles a fleet or an armada, the underlying structure of the LSS must be designed to optimize the balance between strategic coherence and operational flexibility. By doing so, the enterprise can ensure global reach and effective stakeholder engagement, critical components for success in today's complex and fast-paced business environment.

Relating LSS Components

In a large sociotechnical system, the art of relating is central to the architecture's success. Effective coordination among units and between units and external entities is achieved through a web of relationships, each characterized by distinct modes of interaction. These relationships—exchange, organizational, collaborative, cooperative, and compliance—are the conduits through which an LSS operates cohesively within a global context.

CHAPTER 5   ARCHITECTURE

Purpose and Exchange Relations

Each unit's purpose is the beacon that guides its interactions with external stakeholders. The unit's defined purpose magnetizes customers and consumers who resonate with its offerings. Attracted by this purpose, stakeholders connect to the unit, engaging in various transactions such as registrations, sign-ins, and sign-ups. This interaction is governed by exchange relations—coordinated, transactional engagements where offerings are exchanged for value. For instance, a unit dedicated to a customer service platform will see stakeholders engage to receive support, advice, or purchase services, all within a structured and predictable exchange framework. Relations with suppliers, who provide essential resources or services to the enterprise, are also a fundamental type of exchange (transactional) relation, distinct from but equally vital as customer engagements.

Coordination and Collaborative Relations

Collaboration is the heartbeat of inter-unit interactions. Units must synchronize their efforts to ensure that products and services are delivered harmoniously and efficiently. Collaborative relations are established when, for example, the R&D unit develops a new product that the manufacturing unit must then produce. These units must work in concert, sharing insights and aligning timelines to bring the product to market successfully.

Data Sharing and Cooperative Relations

Cooperative relations are established through the sharing of data and information. When one unit possesses information that can benefit another—say, customer feedback that can inform product development—cooperative relations enable the free flow of this data. This cooperation ensures that units are not siloed but rather operate as a network, leveraging shared knowledge for collective benefit.

Direction-Setting, Reporting, and Organizational Relations

Organizational relations define how units within an LSS coordinate to establish accountability and report outcomes. Integrative units, such as a regional headquarter, set strategic directions and boundary conditions, while operational units execute within these parameters and report back on their performance. This establishes a clear accountability process, ensuring that contributions of all units are measured and aligned with the enterprise's goals.

Legitimacy and Compliance Relations

Compliance relations dictate how units interact with regulatory bodies and government entities. These interactions are essential for reinforcing the legitimacy of the enterprise's operations. Whether it's adhering to new environmental regulations or following financial reporting standards, units engage with these external entities to ensure that the enterprise not only meets its legal obligations but also upholds its ethical commitments.

Differentiating and effectively managing these and other relationships (such as investment, employment, community, and so on) enables an LSS to achieve global reach while maintaining a robust operational structure. By prioritizing coordination and clearly delineating the types of relationships, UOEA ensures that units within an LSS can navigate the complexities of global operations, deliver value to stakeholders, and fulfill the enterprise's strategic mission. It's important to note that the nature of these relationships—both among units and between units and external entities—shapes the communication channels, protocols, and types of information exchanges that are feasible and appropriate.

CHAPTER 5   ARCHITECTURE

Benefits of Unit Oriented Enterprise Architecture

The implementation of UOEA presents a multitude of benefits that significantly enhance the operational, structural, and strategic aspects of large sociotechnical systems. Each benefit not only strengthens the individual units within the system but also elevates the LSS as a collective entity. Here, we delve into the top benefits, providing illustrative examples and scenarios.

A Blueprint for a Responsible and Accountable Enterprise

UOEA provides a complete and actionable design that, when implemented, results in an enterprise characterized by its responsible and accountable components.

Example: In a UOEA-implemented healthcare LSS, each unit, be it patient care or medical research, is designed with a clear purpose. While the patient care unit is directly engaged in enhancing patient health and well-being through the application of current medical knowledge and available treatments, the medical research unit operates with a different mandate primarily focusing on the discovery and initial development of new drug compounds. It is dedicated to serving the internal units responsible for drug development and clinical trials. These development units take the medical research outputs as foundational inputs, utilizing this new knowledge to formulate potential pharmaceuticals, advance them through methodical testing stages, and pursue the necessary regulatory clearances. This delineation of purpose ensures that while one unit attends to the immediate health needs of patients, another forges ahead into uncharted scientific territory, ultimately empowering the enterprise to address both current and future health challenges. Through this purposeful design, each unit becomes a responsible element within the larger socially conscious framework of the company, with clear lines of accountability to both internal and external stakeholders.

CHAPTER 5   ARCHITECTURE

Empowerment of Humans via Control over Digital Resources

The architecture harmonizes social structures with digital tools, ensuring that humans remain at the helm of technological resources. Within UOEA, the control over digital resources is comprehensive, extending to the potential for radical alterations or complete overhauls if necessary. This is particularly important given the recent concerns raised by prominent thinkers about the rapid advancements in AI, which have sparked debates around control, ethical usage, and potential unintended consequences. With AI evolving at an exponential rate, there are increasing calls to ensure that human oversight remains paramount, preventing scenarios where technology could operate independently of human intentions or societal values.

Scenario: In a global company, the Regional Sales Unit encounters limitations with their CRM system—it efficiently processes data but falls short in adapting to local market nuances. Empowered by UOEA and within the reasonable constraints set by the regional integrative unit, the sales team identifies and adopts a more agile CRM that incorporates local insights and customizable modules. This shift enables real-time strategy adjustments, aligning technology with market needs and strategic goals. This scenario showcases how UOEA empowers each unit to customize technology for human-driven insights and adaptability, enhancing performance without disrupting other units.

Demystification of Enterprise Architecture

UOEA shifts enterprise architecture from a complex, often opaque discipline to a clear and structured framework focused on the systemic nature of the enterprise. By organizing an LSS into a comprehensible logically structured network of discrete uniformly structured units, UOEA emphasizes clarity of structure, delineates relationships, and highlights

235

CHAPTER 5   ARCHITECTURE

interdependencies. This transparency not only simplifies understanding for stakeholders but also enhances how architects and designers can discuss and communicate the architecture. With a focus on structure and relationships rather than complicated processes, UOEA enables professionals to concentrate on strategic alignment and innovation without the burden of traditional architectural complexity. For example, a fintech enterprise adopts UOEA, restructuring into a highly decoupled, multicloud architecture where each unit operates independently with minimal interdependencies. The flagship unit, or HQ, is deployed on Azure, orchestrating high-level functions and governance, while other units—such as risk analysis, customer accounts, and transaction processing—are hosted on AWS and GCP, each in isolated folders to ensure complete operational autonomy. This extreme decoupling allows units to leverage cloud-native tools suited to their unique functions and regulatory needs without cross-cloud dependencies. The architecture provides resilience and flexibility, enabling each unit to innovate and respond to regulatory changes independently, while Azure HQ maintains overarching alignment across the enterprise.

Transformability, Scalability, and Flexibility of the LSS

UOEA transforms the enterprise into a highly adaptable armada, where entire fleets or individual ships can be added or removed as needed. This modularity allows for the enterprise to respond swiftly to changes in the business environment, market demands, or regulatory requirements.

Take, for instance, an international logistics company operating as an LSS. Under UOEA, each regional division operates like a fleet within an armada. If the company needs to expand its operations to a new geographic market, it can "launch" a new regional fleet rapidly, integrating it into the existing armada without disrupting other fleets. Similarly,

CHAPTER 5 ARCHITECTURE

within each fleet, ships—representing individual distribution centers or services—can be commissioned or decommissioned to scale operations up or down in response to fluctuating logistics demands. Moreover, specific functional units, such as risk and compliance, can be plugged in or out to meet new regulations or standards, thereby ensuring the enterprise remains compliant while continuing its operations with minimal downtime. This level of changeability and scalability is a hallmark of UOEA, offering enterprises unprecedented agility to navigate the complex waters of global business.

Coordination at the Forefront

In UOEA, coordination is no longer an implicit background activity but is instead placed at the forefront of enterprise operations and strategy. By elevating the role of coordination, UOEA ensures that all interactions within the system are intentional, strategic, and aligned with overarching goals. Consider a traditional enterprise where coordination is assumed rather than actively managed. This can result in a disjointed and reactive organization. UOEA challenges this norm by explicitly incorporating coordination channels and protocols into its design, promoting a culture where collaboration and alignment are not left to chance but are embedded in the very fabric of the organization.

For example, in a conventional enterprise, coordination is frequently an underlying function that is not distinctly recognized or designed for. This oversight can lead to inefficient and convoluted communication pathways. UOEA corrects this by consciously designing coordination into the architecture. It emphasizes the significance of indirect coordination mechanisms, such as stigmergic signals, which enable units to coordinate more autonomously based on shared environmental cues and collective intelligence, rather than relying on direct commands. This shift from direct to indirect coordination mirrors the transition from a command-and-control structure to a more agile, responsive, and collaborative network.

CHAPTER 5   ARCHITECTURE

It allows for a fluid and dynamic orchestration of enterprise activities, aligning them more closely with the ever-changing business landscape and stakeholder needs.

Uniformity of Unit Architecture

The consistent standardized structure across units simplifies integration and management.

Scenario: A tech conglomerate operates various business lines from consumer electronics to cloud services. Despite the diversity, each unit's architecture follows a uniform model, with each unit designed as a "ship" uniquely suited to its environment—whether an underwater vessel, an ocean-going ship, or a spaceship. This alignment allows units to operate independently while maintaining a cohesive brand experience and ensuring seamless coordination across the enterprise.

Optimized Communication Networks

Within the framework of Unit Oriented Enterprise Architecture, the extended star topology optimizes communication within a network of units. This topology is essential for eliminating convoluted and inefficient information paths, streamlining the flow of communication across the entire LSS. Building on this, UOEA prescribes a nuanced delineation of communication streams tailored to the nature of relationships between units. It segregates interactions into distinct categories, such as transactional, collaborative, organizational, cooperative, or compliance related, each with a bespoke communication approach. This topology effectively cuts through the clutter of traditional information exchange, paving the way for unobstructed and purposeful communication across the LSS.

CHAPTER 5   ARCHITECTURE

Take, for example, a compliance unit within an LSS. Under UOEA, its role is twofold: externally, it serves as the liaison with regulatory bodies, keeping abreast of regulatory changes; internally, it acts as the central node for gathering compliance reports from other units. This bifurcated approach to communication ensures that regulatory updates received from outside sources are disseminated succinctly to the internal units that require this information to maintain compliance. Concurrently, it consolidates compliance data from within the LSS to report back to external regulators.

This targeted communication strategy facilitated by UOEA allows for a compliance unit to operate with laser focus. It avoids the inefficiency of broad-stroke communications and instead establishes precise channels for information that is pertinent exclusively to compliance matters. The upshot is a communication framework that not only minimizes extraneous interactions but also fortifies the responsiveness and adaptability of an organization to regulatory changes, fostering a state of perpetual readiness.

Case Studies and Applications

Unit Oriented Enterprise Architecture finds its application across various domains where the complexity of large sociotechnical systems can be daunting. By breaking down systems into manageable and coherent units, UOEA facilitates better design, management, and scalability. The following hypothetical case studies illustrate how UOEA can be applied effectively in different scenarios.

Case Study 1: Healthcare Delivery in Smart Cities

> **Background**: A smart city initiative aims to revolutionize healthcare delivery by creating interconnected health units across the city. Each unit comprises a network of medical professionals, patients, digital records, and medical IoT devices.

Application of UOEA: Applying UOEA, each health unit is structured as a self-contained sociotechnical system with a dedicated crew (healthcare professionals) and craft (specialized healthcare management systems). This design enables each unit to operate autonomously, managing patient care and data independently. UOEA facilitates standardized yet secure data interfaces, allowing health units to access patient records relevant to ongoing treatments without centralized data dependencies. IoT devices within each unit monitor patient metrics in real time, integrating with unit-specific systems to track and update patient health data locally. When needed, authorized units can securely access and update patient information, ensuring coordinated and seamless care across the city's healthcare network.

Benefits: The distributed nature of the healthcare units ensures no single point of failure, enhances the capacity for local decision-making, and enables a rapid and targeted response to health crises, like sudden outbreaks, by equipping each unit to act independently yet stay aligned with the city's broader healthcare strategy. Moreover, UOEA also facilitates horizontal scalability, allowing the health system to expand with additional units as the city expands, reinforcing a flexible and responsive healthcare system capable of adapting to growth and future demands.

CHAPTER 5   ARCHITECTURE

Case Study 2: Financial Services Network

Background: A multinational banking corporation seeks to streamline its operations and improve customer service by restructuring its vast array of financial services.

Application of UOEA: Each branch and service department is re-envisioned as a unit with a specific set of purposes, from loan processing to wealth management. The craft includes the digital platforms and tools necessary for each unit's functions. UOEA ensures that these units are interconnected, with clear coordination protocols governing their interactions.

Benefits: The clarity and comprehensibility of the system's digital structure improve significantly, with each unit's purpose and processes clearly defined. The architecture also ensures that changes in one service area can be implemented without disrupting others, resulting in enhanced agility and customer service.

Case Study 3: International Logistics and Shipping

Background: An international logistics company aims to optimize its shipping operations, which have become increasingly complex due to the global scale and the need for real-time tracking and management.

CHAPTER 5   ARCHITECTURE

Application of UOEA: The company's fleet of cargo ships, containers, and logistics hubs are organized into units. Each unit's crew comprises the operational staff, while the craft includes the digital systems for tracking, routing, and inventory management. Coordination is key, with communication standards enabling real-time updates and adjustments across the network.

Benefits: By applying UOEA, the company achieves more efficient routing and better utilization of its fleet. The architecture's emphasis on coordination allows for dynamic reconfiguration in response to changing shipping demands or disruptions, enhancing overall resilience.

Case Study 4: Smart Manufacturing Complex

Background: An industrial conglomerate desires to overhaul its manufacturing operations by implementing smart factories that can adapt to changing market demands and incorporate new technologies.

Application of UOEA: Each smart factory is treated as a unit with its specific production goals and digital augmentation. The UOEA framework integrates production lines (crews) with their associated control systems (crafts), alongside a network of IoT sensors and actuators.

Benefits: The UOEA approach supports a highly adaptable manufacturing environment where

CHAPTER 5   ARCHITECTURE

factories can quickly adjust to new products or processes. The modular structure allows for innovation within units without jeopardizing the stability of the entire manufacturing complex.

Case Study 5: Universal Application of UOEA Across Enterprise Types

Background: Enterprises across the spectrum—be they for-profit businesses, nonprofit organizations, or government entities—are faced with the common challenge of integrating complex human activities with advanced digital technologies to better serve their stakeholders.

Application of UOEA: In this generalized case study, we envision an enterprise that encompasses a range of operations, services, or functions, each constituting a unit within the UOEA framework. The units are designed to be adaptable, with each encompassing a specific subset of the enterprise's overall mission. The crews represent the human element—employees, volunteers, or civil servants—while the crafts are the digital infrastructures that support the unit's operations, such as databases, communication systems, and service delivery platforms.

For-profit implementation: A corporation may have units dedicated to different aspects of its business operations, such as research and development, manufacturing, marketing, and customer service. UOEA facilitates the integration of these units' activities into a cohesive whole, driving efficiency and innovation.

Nonprofit implementation: A nonprofit might organize its units around service delivery, advocacy, research, and fundraising. Each unit within the UOEA framework is clear in its purpose and is equipped with digital tools to maximize its impact on the organization's mission.

Government implementation: Government agencies can structure their various departments and agencies as units. Each unit is focused on a specific aspect of public service, such as public safety, health, education, and infrastructure. The UOEA enables these units to operate with greater autonomy while ensuring that they are still aligned with the overarching goals of the government.

Benefits

UOEA enables these diverse organizations to maintain clarity in their operational structure, facilitating easy navigation of their digital landscape. It encourages uniformity in constructing and managing units, ensuring that each is well defined, purpose driven, and equipped with the appropriate digital tools. For enterprises, this architecture model leads to the following advantages:

- **Clarity in mission and operations**: By defining each unit's purpose and scope, organizations can ensure that every team knows their objectives and how they contribute to the broader goals.

- **Enhanced coordination**: UOEA inherently promotes coordination both within and between units, facilitating a harmonized effort toward the enterprise's goals.

- **Scalability and flexibility**: As new challenges emerge or the enterprise grows, UOEA allows for the addition of new formations, new units, or the reconfiguration of existing ones without disrupting the system's overall integrity.

- **Resource optimization**: By clearly delineating the digital and human components within each unit, enterprises can optimize the use of resources, ensuring that digital tools are effectively supporting human activities.

- **Resilience and adaptability**: In times of crisis or rapid change, the UOEA enables an enterprise to adapt swiftly by modifying or extending individual units without the need for a complete system overhaul.

The case studies presented demonstrate the versatility of UOEA in enhancing the design and function of LSS across various industries. By framing organizations as networks of sociotechnical units, UOEA offers a scalable and resilient framework suitable for modern enterprises facing the challenges of digital transformation and complex operational environments. The adoption of UOEA across various types of enterprises underscores its flexibility and effectiveness in organizing complex systems. By fostering clear, purpose driven design and facilitating the seamless integration of social and digital structures, UOEA enables enterprises to become more agile, resilient, and aligned with their core missions. This architecture model thus stands as a universal approach to system design, capable of enhancing performance and facilitating transformation in a wide array of organizational contexts.

CHAPTER 5   ARCHITECTURE

Integrating Unit Oriented Enterprise Architecture into LSS Strategy

- As technological advancements and global interconnectedness increase the complexity of large sociotechnical systems, the need for a flexible and coherent architectural strategy is essential. UOEA offers a blueprint for designing and managing LSS by emphasizing the strategic alignment of social and digital components.

- **Holistic integration**: UOEA fosters an integrative approach, harmonizing social units with their digital extensions to ensure that technology augments, rather than complicates, human capabilities within each unit.

- **Enhanced accountability**: UOEA empowers each unit with full control over its social and digital extensions, enabling clear accountability for both performance and strategic alignment. This autonomy ensures that each unit is directly responsible for its impact on the broader LSS, fostering a culture of ownership and precision in achieving targeted outcomes.

- **Operational clarity**: By encapsulating functions within distinct units, UOEA provides clear roles and responsibilities, supporting focused, efficient execution while aligning each unit with broader strategic goals.

- **Strategic scalability and adaptability**: Built-in scalability and adaptability allow LSS to expand or adjust quickly to shifts in demand, keeping the system agile and mission focused.

CHAPTER 5   ARCHITECTURE

- **Enhanced autonomy with strategic cohesion**: UOEA empowers units to operate with autonomy, driving innovation and flexibility while ensuring they remain aligned with enterprise-wide objectives.

- **Future readiness**: In an era marked by rapid change, UOEA equips LSS with the flexibility to integrate new technologies and pivot as needed, supporting an enduring competitive edge.

Implementation Imperatives

To successfully integrate UOEA into LSS strategy, leaders should

- Invest in education to build a deep understanding of UOEA principles.
- Foster cross-unit collaboration to leverage the distributed nature of the architecture.
- Align technology investment with the UOEA's modular and interoperable framework.
- Regularly review and refine the interaction patterns to maintain efficient coordination and communication.

Key Takeaways

The adoption of UOEA is not merely the adoption of a new architectural framework; it represents a strategic pivot toward a more resilient, adaptable, and coherent organizational structure. Its principles serve as a beacon for navigating the complexities of modern enterprises, guiding them toward a future where human ingenuity and digital potential are

CHAPTER 5 ARCHITECTURE

inextricably linked. In a world that changes every instant, UOEA offers a methodology not just to survive but to thrive amidst the flux, making it an indispensable element of any forward-thinking LSS strategy.

Here are the key takeaways from the chapter, highlighting the foundational principles and strategic methodologies of Unit Oriented Enterprise Architecture and its application to large sociotechnical systems:

- **Influence of the Fleet metaphor**: Unit Oriented Enterprise Architecture is deeply influenced by the philosophy and attributes integral to the Fleet metaphor.

- **Comprehensibility and changeability**: Central to this architectural strategy are two pivotal concepts—the comprehensibility of the architecture itself and the changeability of the LSS as a whole.

- **Three essential processes**: In shaping the architecture of LSS, we embrace a simple methodology that comprises three essential processes: selecting components, positioning components, and relating components.

- **Uniformity in component selection**: We favor uniformity in selecting high-level LSS components, aiming to enhance interoperability, comprehensibility, and clarity within the system.

- **Extended Star pattern**: In architecting LSS, the Extended Star pattern ensures harmony between the core and peripheral aspects, maintaining strategic oversight while enabling responsiveness and adaptability.

CHAPTER 5   ARCHITECTURE

- **Relational patterns**: The primary relational patterns are exchange, organizational, collaboration, cooperation, and compliance relations between LSS components and external entities.

- **Compositional patterns**: The selection of compositional patterns is closely tied to the levels and goals of the system, ranging from Distributed at the Fleet level and Partitioned at the Ship level to Interlocked and Fused at the more granular levels.

- **Enterprise prototype**: We select a large enterprise with a geographical divisional structure as a prototype for analysis and modeling, shaping our approach to designing large-scale sophisticated systems.

- **Blueprint for accountable enterprise**: UOEA provides a blueprint for a responsible and accountable enterprise, aligning social and digital architecture, ensuring clear ownership, empowering humans, enhancing clarity, and streamlining communication flows.

CHAPTER 6

Design

> *With the precision of a poet, the author weaves together Purpose, Function, Process, Structure, Order, Culture, Memory, and Storage—words once thought to be at odds, now harmoniously aligned in the mosaic of his innovative design philosophy. And just when you thought the jigsaw was complete, the author unveils the pièce de résistance, Purpose Driven Design—because what's a magnum opus without that dash of je ne sais quoi that invites you to marvel at the audacity of its brilliance?*

The initial chapters have established a groundwork by presenting the unit oriented enterprise architecture—a novel architectural approach that organizes large sociotechnical systems into fleets of sociotechnical units, each embodying a harmonious fusion of human skills and technological competence. This architecture accentuates the systemic characteristics of sociotechnical systems, the requirement for understandability, and the paramount significance of accountability. It ensures that each unit functions within a clearly delineated range of responsibilities and goals, thereby optimizing its contribution to the overall success. Let's now consider Purpose Driven Design (PDD) of sociotechnical units and design principles that facilitate the seamless integration of human and technological components.

CHAPTER 6   DESIGN

Figure 6-1. *In Evolveburgh, the new "fleet-themed" by-law has residents out with signs like "We Want Sidewalks, Not Ship Decks!", "Homes, Not Hulls!", and "Keep Our Homes Above Sea Level!"—highlighting the absurdity of underwater markets and space shuttle suburbs*

Designing Units: Balancing Purpose-Driven Logic and Structural Uniformity

In the architectural vision of UOEA, the design of units within an LSS is approached from two distinct but complementary angles. This bifocal approach is fundamental to achieving a balanced and effective LSS where uniformity facilitates coherence and integration, while purpose-driven logic ensures that each unit is uniquely equipped to fulfill its specific role.

- **Structural uniformity across the LSS**: On one side of the design spectrum, there is a drive for uniformity. All units across the LSS share a common structural blueprint: they are composed of digital tools that include apps, applications, and suites of applications. This standardized structure is deliberate, fostering a seamless operational environment where each unit, regardless of its unique purpose, upholds the system-wide coherence. It simplifies numerous aspects of system management, from technical support to the scalability of operations, by ensuring that all units adhere to a foundational architectural design.

- **Purpose-driven logic**: On the other side, there is the principle of purpose driven design, which tailors each unit to meet specific goals and stakeholder needs. While the uniformity of structure provides a shared skeleton, the flesh and muscle of each unit are formed according to the particular functions it must perform within the LSS. UOEA integrates purpose, function, structure, order, culture, memory, and storage constructs into a coherent digital chain to equip each unit to pursue its purpose effectively. The fusion of these constructs ensures that the digital environment of the unit is not just a collection of tools but a living ecosystem that facilitates the unit's mission. This angle of design recognizes that each unit has a unique value proposition and role to play in the broader ecosystem of the enterprise.

CHAPTER 6   DESIGN

This dual design approach in UOEA is pivotal for creating an LSS that is robust yet flexible, standardized yet adaptable. It is a strategic acknowledgment that a successful enterprise must not only maintain order and efficiency through uniform architecture but also embrace the diversity of purpose that comes with its varied units.

Structural Uniformity

The composition of units within a large sociotechnical system often faces a challenge common in the industry: the confusion stemming from an array of technological concepts. Architects and engineers are frequently bombarded with jargon—applications, microservices, workloads, and more—each carrying its weight and significance, yet seldom is there clear guidance on synthesizing these elements into a unified whole. This lack of clarity can lead to fragmented systems that lack harmony and coherence.

Simplifying the Building Blocks: Applications and Suites

To cut through this confusion, UOEA simplifies the approach by proposing suites of applications as the primary and singular type of building block for composing a "ship" within the LSS "fleet." Each suite consists of applications, and each application is made up of apps (microservices). This hierarchical structure is justified by the need for clarity and the efficacy of modular design.

CHAPTER 6   DESIGN

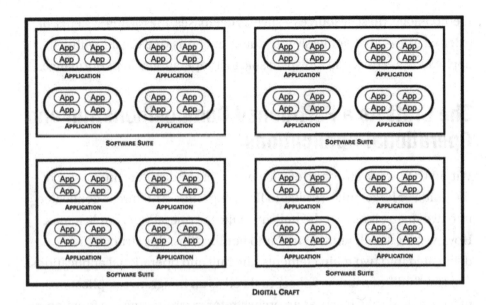

Figure 6-2. *Structural uniformity of a sociotechnical unit*

Just as a ship is composed of various modules—navigation, propulsion, and life-support systems, to name a few—each designed to perform specific functions while being part of the greater vessel, suites within a unit serve analogous roles. They are independent in functionality yet integrate seamlessly to form a coherent unit. A suite of applications can be likened to a "zone" on a ship, such as the bridge or engine room, where a collection of systems operates in concert to fulfill broader operational objectives.

Backing Services As Extensions

Within a unit, additional resources are not merely considered extensions but are intrinsic parts of the applications themselves, functioning as essential components that enhance and complete the capabilities of the main building blocks. Following the principles outlined in the Twelve-Factor App methodology, these resources—such as databases, messaging systems, or external APIs—are seamlessly integrated, acting as vital organs within the application's ecosystem, indispensable to its operation and performance.

The Craft As a Partitioned Composition: Hull and Operational Applications

Within UOEA, the structure of each unit's "craft" depends on the size of the unit. For large units, similar to fleet ships, this structure is delineated into two main elements: the hull and suites of operational applications. However, for smaller units, similar to flotilla ships, this structure is delineated into two main elements: the hull and operational applications.

The "hull" of a unit is conceptualized as an integrative application, playing a dual role. It acts as an orchestrator, coordinating the functioning of other applications and ensuring they operate in sync with each other. Simultaneously, it serves as the unit's facade, the interface through which the unit interacts with the rest of the LSS and the external environment. This dual functionality is critical for maintaining both the internal integrity and external connectivity of the unit.

Operational applications are similar to the various functional components of a ship, such as the navigation system, cargo holds, or living quarters, each serving a specific operational need but collectively forming a cohesive entity. These applications are integrated into the hull, which coordinates their activities and serves as the point of integration that ties the individual operational elements to the larger LSS.

Integration of Hull and Operational Applications

The integration of the hull with operational applications is a carefully orchestrated process. It involves establishing a cohesive and interoperable system architecture where the hull application provides foundational services that support the operational applications. These might include user authentication, data processing pipelines, or shared utility services that are fundamental to the operation of the individual applications.

This configuration ensures that the craft operates as a singular, harmonious entity. The hull, serving as the integrative application, not only aligns and unites the operational applications but also provides a stable platform upon which they can function effectively. Through this symbiotic relationship, the craft maintains the operational coherence and strategic alignment of the unit, much like the architecture of the LSS which is built upon the integrative unit and operational units within the extended star topology framework.

Digital Chain of Purpose-Driven Constructs

In the process of designing units within an LSS, the second key aspect is the formation of a "construct chain." This chain, driven by a well-defined purpose, acts as the unit's structural backbone. It orchestrates the integration of diverse digital and social components, thereby ensuring the unit's mission is accomplished effectively.

CHAPTER 6   DESIGN

Figure 6-3. *The Purpose Driven Design framework*

The construct chain is an arranged sequence of digital constructs, encompassing

- **Purpose**: The fundamental rationale behind the unit's existence

- **Function**: The interfaces that define the unit's interactions and operations

- **Process**: The step-by-step actions undertaken by the unit

- **Structure**: The organizational layout or arrangement of the unit

CHAPTER 6   DESIGN

- **Order**: The governance, rules, and constraints that provide a systematic framework
- **Culture**: The collective attitudes, values, goals, and practices prevalent within the unit
- **Memory**: The unit's ability to retain and recall information
- **Storage**: The system(s) dedicated to storing digital products

Each element in this chain is strategically aligned and seamlessly integrated, ensuring a harmonious operation of the unit's various aspects.

Purpose Components and Identity Manifestation

The purpose constructs are the soul of the unit, manifesting its identity and intentions. They are typically implemented as the unit's marketing facade, which includes elements such as the website's entry page, microsites, landing pages, and social media profiles. These components serve as beacons that attract and engage customers or consumers, reflecting the unit's raison d'être and value proposition.

Function Constructs As Interaction Facades

Function constructs are implemented as the unit's main interaction facade. They offer consumers multiple channels—web, mobile, conversational interfaces, APIs, and IoT devices—through which they can engage with the unit. These channels facilitate a host of actions, such as registration, sign-in, sign-up for offerings, purchasing products, and utilizing services. They are the direct touchpoints where consumers interact with the unit, securing its operational relevance.

CHAPTER 6   DESIGN

Process Constructs: The Digital Conveyors

Process constructs are the digital equipment, mechanisms, and tools that underpin the function by aiding coordination and execution. These constructs provide the necessary digital infrastructure for both human and artificial agents to perform work and produce the unit's outputs. They ensure that the workflows are efficient, coherent, and aligned with the unit's objectives.

Structure Constructs: The Crew Framework

Structure constructs are concerned with the organization of the unit's social component—the crew. This includes the assembly of purposeful, responsible, and accountable teams that perform work to support processes. The arrangement of these teams is carefully designed to foster effective collaboration and to drive the unit's processes forward.

Order Components and Regulatory Manifestation

The order constructs represent the unit's regulatory framework, manifesting its discipline and control mechanisms. They are typically implemented as the unit's governance models, rule sets, and constraints. These components serve as the guiding principles that shape and direct the unit's actions, reflecting the unit's commitment to systematic and orderly operations. They ensure that every process within the unit adheres to a set of standards, thereby maintaining consistency and efficiency in the unit's operations. This manifestation of order articulates the unit's dedication to a well-regulated and controlled approach, underlining its value proposition.

CHAPTER 6   DESIGN

Culture Constructs: The Repository of Values

Culture constructs hold the unit's cultural artifacts and collective knowledge, encompassing values, myths, beliefs, taboos, rituals, symbols, and reward systems. These constructs are typically housed in a content management system and play an important role in creating a positive and productive work climate.

Memory: The Data Chronicle

Memory constructs encompass the relational and NoSQL databases that record operational data, decisions, events, processes, and tasks. They serve as the unit's historical ledger, enabling analytics, supporting continuity, and providing an audit trail. This data chronicle is essential for maintaining a record of the unit's activities, informing strategic decisions, and validating the unit's performance.

Storage Constructs: Digital Warehousing

Storage constructs utilize digital storage solutions to house the unit's outputs—products and services ready for deployment through function. This digital warehousing enables the unit to manage its inventory efficiently, ensuring that its outputs are secure, accessible, and primed for distribution.

This detailed and interconnected digital chain of constructs ensures that the unit's purpose is reflected at every touchpoint and that its operations are executed with precision and alignment, thus fulfilling the unit's mission while upholding the principles of structural uniformity across the LSS.

Through this two-part approach—ensuring structural uniformity while embedding purpose-driven structural logic into each unit—UOEA guides the composition of LSS with a clarity and precision that fosters robustness, scalability, and efficiency. It provides architects and engineers with a clear path to designing and implementing units that are both independent in capability and integrated in purpose, forming the resilient and responsive vessels of the enterprise's global fleet.

Purpose Driven Design: Core Principles

The core principles of unit design in a large sociotechnical system revolve around purposefulness, modularity, clarity, and integrative functionality. By structuring units into suites of applications and further into apps (microservices), each with a designated integrative component, these principles ensure coherence, seamless integration, and efficient operation within the larger system.

Principle 1: Purpose Driven Sociotechnical Wholeness

What: Each system within an LSS represents a unified whole aligned with a specific purpose.

Why: By focusing on purpose driven design, this principle ensures that every component within the system contributes to a coherent purpose, facilitating a clear and directed orchestration of both social and digital activities. Ultimately, purpose driven design provides a clear path for each unit to contribute meaningfully to the LSS's overarching objectives, ensuring that the system's complexity does not hinder but rather enhances its capability to adapt and thrive in dynamic environments.

CHAPTER 6   DESIGN

Principle 2: Coherence in Purpose Implementation

What: Each sociotechnical unit should exhibit a coherent implementation chain that includes digital function, process, structure, order, culture, memory, and storage, all geared toward fulfilling the unit's purpose.

Why: This principle highlights the importance of a well-synchronized operational system where every digital and social component is methodically aligned to serve the unit's purpose. Such coherence ensures that the functional activities, procedural flows, structural arrangements, rules and constraints, cultural dynamics, and informational continuity collectively translate the unit's overarching goals into tangible outcomes. It is this purpose-driven coherence that transforms a mere collection of elements into a synergistic and functional whole, reinforcing the unit's role within the larger sociotechnical system. The term "implementation" in the title encapsulates the active, dynamic process of turning strategy into action, where the focus is on the actualization of objectives through a unified and supportive framework of system components.

CHAPTER 6   DESIGN

Principle 3: Uniformity and Accountability in Sociotechnical Units

What: Each sociotechnical unit within an LSS must be designed with uniform structure, maintaining legitimacy, purposefulness, and accountability, while the overall composition of the LSS adheres to a distributed architecture style.

Why: The uniformity in unit design ensures coherence and compatibility across the LSS, streamlining communication and collaboration. Legitimacy ensures that each unit has a clear and recognized role within the system, purposefulness ensures that each unit is designed with intent and direction, and accountability means that units are responsible for their performance and outcomes. The application of a distributed architecture at the macro level of the LSS supports decentralized control and facilitates transformability and scalability, with each unit operating as an independent yet integrated part of the whole. This approach enhances the system's ability to adapt to change and distribute operational loads effectively.

Principle 4: Autonomy and Control Within Sociotechnical Units

What: The social constituents within a sociotechnical unit should have sufficient autonomy and full control over their digital extensions.

Why: Autonomy empowers units to respond swiftly to changes and challenges, while control—interpreted as legitimate possession—ensures that units can steward their resources effectively without the encumbrances of ownership. This dynamic promotes a sense of responsibility and fosters an environment where resources can be optimized for the unit's needs.

Principle 5: Composition of Sociotechnical Units

What: Sociotechnical units are composed of a crew (social component), a digital craft (primary digital extension), and remote digital objects like smart objects or other remote digital structures.

Why: This composition mirrors the operational needs of real-world vessels, where the crew is supported by a craft that facilitates their mission. The inclusion of remote digital objects expands the unit's operational reach and capability, integrating the Internet of Things to enhance functionality and situational awareness.

Principle 6: Purpose-Driven Digital Craft

What: A digital craft within a unit consists of various constructs—purpose, function, process, structure, order, culture, memory, and storage—that support the unit's purpose and deliver value through products or services.

CHAPTER 6   DESIGN

Why: By organizing the digital craft into purpose-supporting constructs like function, process, structure, order, culture, memory, and storage, each unit becomes a microcosm of value delivery. This organization ensures that every technological aspect is tuned to serve the unit's goals, fostering a direct correlation between the unit's purpose and its operational output.

Principle 7: Modular Composition of Digital Crafts

What: Digital crafts are composed of suites of applications, where each suite consists of applications, and each application is made up of apps (microservices). Each level of this hierarchical structure is networked and systemic, meaning there is an integrative suite, application, or app responsible for coordinating the others and providing public APIs and/or user interfaces.

Why: This systemic structure ensures modularity, clarity, and efficient operation within the LSS. An integrative component at each level coordinates the functions of the other components, thereby maintaining coherence and seamless integration. This approach helps prevent fragmentation and confusion, enabling the system to function as a unified whole while allowing for scalable and flexible development and maintenance.

CHAPTER 6   DESIGN

Key Takeaways

Here are the key takeaways from the chapter, highlighting the dual approach of uniformity and purpose-driven logic in the design of units within large sociotechnical systems. These points emphasize the importance of a common structural blueprint, networked and systemic structures, and the integration of specific goals and stakeholder needs to create a balanced and effective LSS.

- **Bifocal design approach**: In the architectural vision of UOEA, the design of units within an LSS is approached from two distinct but complementary angles, achieving a balanced and effective system through uniformity and purpose-driven logic.

- **Structural uniformity**: Structural uniformity across the LSS is emphasized, with all units sharing a common structural blueprint composed of digital tools, including apps, applications, and suites of applications.

- **Networked and systemic structures**: A unit of a large sociotechnical system (LSS) consists of suites of applications, each suite consists of applications, and each application consists of apps (microservices), with one integrative component at each level coordinating others and providing public APIs and/or user interfaces.

- **Purpose-driven logic**: The principle of Purpose Driven Design tailors each unit to meet specific goals and stakeholder needs, integrating purpose, function, structure, order, culture, memory, and storage constructs into a coherent digital chain.

CHAPTER 6   DESIGN

- **Seven core principles**: The seven core principles of Purpose Driven Design in a large sociotechnical system revolve around purposefulness, modularity, clarity, and integrative functionality.

CHAPTER 7

Implementation

> *The author endeavors to convince us that Implementation is anchored by the twin pillars of Organization and Engineering, which together construct a pathway to success—a journey demanding not just diligence and resilience, but also a touch of wizardry. Just be mindful that this path, paved with the yellow bricks of overtime and brown rivers of espresso, may lead you not just to the Wonderful Wizard of Oz but to his less-talked-about cousin, the Wizard of Ops, who, instead of granting wishes, will most likely ask you to reboot the server and check the error logs. Good luck on your quest!*

In the journey of creating and maintaining large sociotechnical systems, engineering plays a pivotal role in turning architectural blueprints into tangible, functioning entities. This chapter begins by bridging the gap between the theoretical constructs of Unit Oriented Enterprise Architecture and the practical aspects of bringing these concepts to life. LSS engineering is not merely about the technical assembly of system components; it encompasses a holistic approach that integrates the complex interplay of technology, human factors, and organizational dynamics.

The transition from architectural design to implementation in LSS engineering is nuanced and multifaceted. It requires a deep understanding of the architectural principles outlined in UOEA, coupled with practical skills in system design, technology integration, and process management. This phase is important, as it involves translating the high-level designs and strategic visions into actionable plans and concrete systems.

CHAPTER 7   IMPLEMENTATION

As we explore the implementation of LSS, we discuss at a high level how the abstract concepts of structure, purpose, and relational dynamics within an LSS materialize into real-world software systems. This process entails a sequence of strategic choices, from selecting the appropriate technology stack and designing scalable data structures to safeguarding system security and ensuring compliance with regulatory standards. However, these detailed aspects are beyond the scope of this book, and our focus will remain on the overarching considerations for LSS implementation.

Figure 7-1. In Evolveburgh, they've divided implementation into two "critical" mechanisms: Organization and Engineering. It's like saying you need both bread AND butter for a toast. A toast to the implementers of Evolveburgh: May your toasts always land sunny-side up and your projects never get "jammed."

The implementation of LSS is fundamentally iterative and adaptive. The social dimension of LSS, which encompasses the living entities that evolve over time, underscores the importance of code organization in the implementation process. This aspect is frequently overlooked, not just by engineers but also by major software vendors. Consequently, there is a need for implementation strategies that are not only robust but also sufficiently flexible to accommodate shifts in business environments, technological progress, and the changing needs of users.

CHAPTER 7　IMPLEMENTATION

By distinguishing between code organization and engineering, practitioners can successfully close the gap between theoretical concepts and practical application. This approach enables the creation of systems that are not only technologically robust but also socially pertinent and organizationally efficient.

Figure 7-2. In Evolveburgh's latest feat of bureaucratic brilliance, city officials cut a ribbon across the symbolic "Theory-Practice" gap, proudly declaring it "Closed by Executive Order"—proving that in this town, all it takes to bridge complex issues is a big stamp, a little red tape, and a hefty dose of wishful thinking

CHAPTER 7   IMPLEMENTATION

Translating UOEA Principles into Implementation Practices

Two mechanisms—Organization and Engineering—play a pivotal role in transforming the theoretical constructs of UOEA into concrete, functional systems.

Organization: Distributed and Partitioned Composition

The first mechanism, *Organization*, centers around the deployment of the distributed composition of the enterprise. As a distributed system, a large sociotechnical system is a collection of independent units located on different cloud regions or even different clouds. These units coordinate and communicate with each other, ensuring that despite their geographical dispersion and operational independence, they operate cohesively as a unified whole. This approach underlines the essence of distributed architecture in UOEA, where decentralization is key to both resilience and strategic flexibility. It is similar to setting up a series of interconnected but autonomous operational hubs, each functioning within its own cloud or multicloud environment. This setup not only provides resilience and scalability but also ensures that each regional unit can respond effectively to local market conditions and regulatory environments.

At the unit level, the principle of partitioned composition comes into play. This involves organizing each unit within its designated cloud environment(s). The partitioned approach means that while each unit operates as part of the larger enterprise ecosystem, it maintains its distinct operational boundaries and capabilities. This structure allows for a high degree of specialization and customization within units, enabling them to cater to specific functional requirements or stakeholder needs while still adhering to the overarching enterprise strategy.

CHAPTER 7 IMPLEMENTATION

Figure 7-3. *Evolveburgh's auditors have uncovered a "cloudy" trend in code organization, where enterprise code is more scattered than a teenager's attention span during a history lecture. Developers are now sending "wish you were here" postcards to their lost snippets, attaching GPS trackers in hopes they'll find their way back. The Chief Engineer became Evolveburgh's Chief Meteorologist, forecasting code showers and scattered functions with a chance of recursive rain. Locals now carry umbrellas to dodge the downpour of bits and bytes. Some have upgraded to hard hats because you never know when a wild loop might drop from the sky*

CHAPTER 7 IMPLEMENTATION

Engineering: Development and Integration of Applications

The second mechanism, *Engineering*, delves into the practical aspects of software development and integration of a unit's building blocks—the applications. The engineering process in UOEA involves developing applications that align with the unit's defined purpose and function. This task requires a deep understanding of both the technical aspects of software development and the operational context in which these applications will function.

CHAPTER 7 IMPLEMENTATION

Figure 7-4. *Welcome to Evolveburgh's Engineering Antipatterns Museum, where the code is so wild, we've upgraded from bug spray to hazmat suits! In our Debugging Horrors Hall, you'll find bugs with talents that would make any circus performer jealous. Play in the Gargantuan Glitches Playground, where "try/catch" is not a game, but a way of life. Marvel at the Exhibit of Exceptional Errors, where "404 – Brain Not Found" is the most popular piece. Wander through the Repository of Recursive Regrets, but be careful not to get stuck in an infinite loop of "what ifs." Visit the Unused Code gallery, where "dead code" takes on a whole new meaning. And don't miss Dependency Hell—it's hotter than your laptop after an all-night coding session. We've got aloe vera on hand, because we know that feeling of getting burned by a stack overflow. Hold your nose for the Code Smell exhibit; it's not your average stink. We're talking about a fragrance that could knock out a skunk! Remember to swing by our gift shop for a souvenir gas mask. Trust us, it's the latest fashion statement in programmer's gear—because let's face it, your codebase probably has its own peculiar perfume too!*

CHAPTER 7 IMPLEMENTATION

The development process is typically agile, focusing on creating modular, scalable, and maintainable applications. These applications are designed to operate seamlessly within the cloud environment of the unit, leveraging cloud-native technologies and practices for optimal performance and resilience.

Integration is another critical aspect of the engineering mechanism. Once developed, these applications must be integrated not just within the unit but also with other units and external entities. This broader scope of integration ensures seamless interaction and data flow across the LSS, enhancing the system's overall functionality and its ability to interact effectively with the external environment, including customers, partners, and regulatory bodies. This integration is facilitated through well-defined APIs, microservices architecture, and event-driven mechanisms, ensuring that data and functionalities can flow smoothly across the enterprise's digital ecosystem.

In summary, translating UOEA principles into implementation practice within a large enterprise involves a twofold approach. The *Organization* mechanism ensures that the enterprise is structured in a way that is both distributed and partitioned, aligning with the cloud-centric, geographically diverse nature of modern business operations. The *Engineering* mechanism, on the other hand, focuses on the nuts and bolts of building and integrating the digital components of each unit, ensuring that they not only fulfill their specific roles but also contribute to the enterprise's holistic goals. This dual approach ensures that the principles of UOEA are effectively grounded in the practical realities of running and managing a large-scale, technologically sophisticated enterprise.

CHAPTER 7   IMPLEMENTATION

Organizing LSS and Units

Organizing the LSS and individual units within cloud environments is a strategic endeavor that significantly influences the system's efficiency, compliance, and functionality. Various strategies can be employed to structure LSS and units effectively within cloud environments, leveraging the benefits of both single-cloud and multicloud configurations. If the enterprise uses multicloud strategies, special attention must be given to the complexities of managing multiple cloud environments, the advantages in terms of resilience and flexibility, and the considerations for integration, data management, and consistent policy enforcement across diverse cloud platforms.

By taking into consideration geographical, departmental, environmental, and project-based factors, along with a structured approach to naming and tagging, enterprises can build a cloud infrastructure that is agile, compliant, and aligned with their strategic objectives.

Geographical Organization

Aligning cloud resources with the enterprise's geographical divisions is essential. This method involves structuring resources based on geographical regions, which has several advantages:

- Making compliance with regional data laws more manageable, as data storage and processing can be localized according to legal requirements.

- Improving service latency for local users by hosting resources closer to where they are accessed.

CHAPTER 7 IMPLEMENTATION

Departmental Segmentation

Organizing cloud resources according to different departments or business units such as HR, Finance, and Marketing ensures efficient resource utilization:

- Facilitating managing access controls, allowing departments to have autonomy over their resources while maintaining overall security
- Streamlining resource allocation, catering to specific departmental needs and workflows

Environment-Based Segmentation

Segmenting resources for different operational environments like development, testing, and production is essential to maintain system integrity:

- Preventing conflicts between development stages, ensuring that testing or development activities do not impact production environments
- Supporting smooth deployment cycles and continuous integration/continuous deployment processes

Project- or Service-Based Structure

For enterprises with multiple projects or services, organizing resources per project or service offers operational clarity:

- Enabling precise tracking of costs and performance metrics for each project or service, aiding in resource optimization and budgeting
- Facilitating project-specific customization and scalability, adapting resources to the unique demands of each project

Effective Naming Conventions

Developing and implementing effective naming conventions is vital for resource organization and management:

- A consistent naming scheme that incorporates elements like geographical location, department, project, and environment in resource names enhances clarity.

- Hierarchical naming for dependent or related resources simplifies understanding their interrelationships.

- A robust tagging strategy categorizes resources by various parameters like ownership, purpose, confidentiality, or cost center, enriching resource metadata for better management.

These organizing activities empower unit crews with full control over their digital resources, enabling them to effectively manage

- **Access control and permissions**: Implement geographical and departmental permissions and role-based access control to enhance security and operational efficiency.

- **Cost allocation and tracking**: Facilitate departmental cost tracking and geographical cost analysis, providing insights for budget optimization and resource allocation.

- **Compliance and data sovereignty**: Enforce regional compliance policies and data localization as required by law or for performance, ensuring adherence to regional data protection regulations.

- **Service catalogs and standardization**: Develop standardized service catalogs for commonly used resources and configurations to streamline deployment across departments and regions. Utilize templates and blueprints for common infrastructure setups to ensure uniformity and adherence to organizational standards.

Multicloud Approaches in LSS Organization

When organizing large sociotechnical systems within cloud environments, adopting a multicloud approach presents unique challenges and opportunities. Multicloud strategies offer significant advantages in terms of resilience and flexibility:

- **Enhanced resilience**: By distributing resources across multiple clouds, LSS can achieve higher fault tolerance and reduce the risk of downtime.
- **Greater flexibility**: Multicloud environments allow organizations to choose the best services from different providers, tailoring the cloud setup to specific operational requirements and optimizing performance.

However, the management of multicloud environments is inherently complex due to the involvement of different cloud service providers, each with their unique features and operational models. Navigating these complexities requires:

- **Robust coordination**: Ensuring effective communication and synchronization between different cloud environments to maintain operational consistency.
- **Expertise in diverse platforms**: Developing an in-depth understanding of each cloud platform's specific tools, services, and management practices

CHAPTER 7 IMPLEMENTATION

- **Unified management tools**: Leveraging advanced cloud management solutions that provide a centralized view and control over different cloud resources

Figure 7-5. *The city manager of Evolveburgh scolds his architects for keeping the multicloud concept under wraps, exclaiming, "Why settle for one cloud to hide spending when we could triple the fun?" Meanwhile, the architects clutch blueprints, realizing they may have just invented the world's first taxpayer-funded fog machine*

CHAPTER 7   IMPLEMENTATION

AWS Implementation for LSS

In the context of AWS, aligning with its recommended best practices, it's advisable to create a single organization under AWS and manage all accounts within this organizational structure. This organization acts as a security and policy boundary, allowing for consistency and centralized control across the LSS environment.

For an LSS following the UOEA principles, each unit should be structured as an individual AWS account within this organization. This approach offers several benefits:

- **Security and isolation**: Each unit's cloud resources and data are contained within its own account, providing a strong identity and access management isolation boundary. This structure enhances security as it, by default, prevents access between units (accounts), ensuring data and resource integrity.

- **Operational excellence and flexibility**: Using multiple accounts enables tailored resource management and access controls for each unit. It allows for operational practices to be optimized according to the specific needs of each unit, aligning with AWS's Well-Architected Framework pillars including operational excellence and reliability.

- **Centralized governance and consistency**: Within the single AWS organization, policies and service-level configurations can be centrally applied across all unit accounts. This setup facilitates consistent policy enforcement, central visibility, and programmatic controls across the multi-account LSS environment, contributing to overall governance and cost optimization.

CHAPTER 7   IMPLEMENTATION

- **Resource sharing and inter-unit collaboration:**
  While each account provides isolation, AWS also offers mechanisms to explicitly allow resource and data sharing between accounts when necessary. This capability is vital for inter-unit collaboration and resource optimization within the LSS, allowing units to share resources in a controlled and secure manner.

In summary, organizing each unit of an LSS as a separate AWS account within a single AWS organization aligns with AWS's recommendations and enhances the overall security, governance, and operational efficiency of the LSS. This structure allows for the unique needs of each unit to be met, while maintaining a cohesive and controlled environment at the organizational level.

Azure Implementation for LSS

When implementing an LSS in Azure, organizations are presented with the option of either a single tenant or a multitenant structure. While Microsoft typically recommends adopting a single tenant structure for each organization, the choice largely depends on a variety of factors specific to the company. These can include organizational, strategic, and even political considerations, each playing a significant role in determining the most suitable architectural approach for the organization's unique needs and objectives.

Organizations may choose a multitenant structure in Azure for various strategic reasons. Conglomerates with multiple subsidiaries or business units that operate independently often find a multitenant setup more conducive to maintaining operational autonomy. This approach also aligns well with organizations undergoing mergers and acquisitions, as it simplifies the integration of diverse IT systems and business processes. Additionally, in situations involving divestiture activities where a business is split to form or merge with another entity, separate tenants

283

CHAPTER 7   IMPLEMENTATION

can facilitate smoother transitions and clearer operational segregation. Multitenant structures are also advantageous for organizations that need to exist in multiple cloud environments due to compliance or regulatory requirements or for those operating across multiple geographic locations with different residency regulations. Each of these scenarios presents unique operational and governance challenges that can be more effectively managed through a multitenant architecture in Azure.

Regardless of whether a single tenant or multitenant approach is chosen, Azure facilitates the creation of a flexible structure of management groups and subscriptions. This hierarchy allows for unified policy and access management across the LSS.

In a **single tenant structure**, one can establish a Root Management Group for the entire enterprise, creating an overarching governance scope. Under this Root Management Group, individual management groups for each business unit can be established. This approach offers unified governance across the enterprise, with policies and access management uniformly applied.

In a **multitenant structure**, separate tenant Root Management Groups can be created for each geographical region or subsidiary. Within each Root Management Group, management groups for each business unit provide localized governance. This setup allows for more tailored policy and access management, catering to the specific needs and regulations of each region or subsidiary.

A single tenant structure creates an opportunity for centralized governance, simplifying policy and access management across the entire LSS. This is particularly beneficial for enterprises seeking uniformity and consistency in policy enforcement.

A multitenant structure offers more flexibility, especially for regional governance scenarios. It allows for policies and access controls to be customized according to the specific needs of different regions or business entities, enhancing the adaptability of the LSS to varied operational and regulatory environments.

CHAPTER 7   IMPLEMENTATION

The choice between single and multitenant structures in Azure for LSS implementation should be based on the specific needs and characteristics of the organization. Both approaches offer robust mechanisms for organizing resources and managing governance, but they cater to different operational and administrative requirements. By carefully evaluating their unique circumstances, enterprises can choose the most suitable structure to optimize their LSS in Azure.

GCP Implementation for LSS

When structuring an LSS in Google Cloud Platform (GCP), organizations can leverage its straightforward yet effective resource hierarchy, which comprises four main layers: Organization, Folders, Projects, and Resources. This hierarchy is designed to provide comprehensive control and organization of resources, aligning with the needs of an LSS.

At the top of the hierarchy is the Organization (Org Node), representing the highest level within GCP's structure. The Organization layer symbolizes the company itself and is pivotal for centralized administration. It provides overarching control over critical aspects like billing, resource management, and access policies.

Within an LSS framework, the Org Node acts as the central point for managing all cloud resources across the enterprise, ensuring unified governance and oversight.

The next layer in the hierarchy comprises Folders. Folders in GCP are instrumental in grouping resources that share common Identity and Access Management (IAM) policies. Each Folder can encompass multiple subfolders or resources but maintains a singular parent, ensuring a clear and manageable organizational structure. Folders are exceptionally versatile and can be utilized to represent both divisions and departments/business units within a company. This allows for a flexible yet organized approach to structuring the LSS. For instance, a Folder could represent a specific geographical division like "European Operations" or a department

285

such as "Marketing." Subfolders within these can further delineate the structure, reflecting the internal organization of these divisions or departments. Policies applied at the Folder level cascade to all included resources, enabling efficient governance and policy enforcement across different sections of the LSS.

By using Folders to mirror the organizational divisions and departments, GCP allows for a structured yet adaptable LSS environment. This setup aligns with the principles of a flat organizational structure, as advocated by UOEA, while providing the flexibility to adapt to the specific operational and governance needs of each unit.

The ability to apply policies at different Folder levels also aligns with the need for differentiated governance across various parts of the LSS, catering to specific departmental or divisional requirements while maintaining a cohesive overall management system.

In summary, GCP's simple yet effective four-layer hierarchy is well suited for organizing and managing resources in an LSS. The Organization layer ensures centralized control and oversight, while the Folders layer offers a flexible means to structure the system according to the company's divisions and departments, facilitating effective governance and resource organization within the LSS.

Engineering Units

In UOEA, engineering sociotechnical units involves a detailed focus on their core components: applications and suites of applications. These elements are uniformly structured, forming the fundamental building blocks of each unit within an LSS.

Uniform Structure of Units

Each unit in UOEA is constituted either of stand-alone applications or by cohesive suites of applications. This uniformity in structure is essential for ensuring consistency, interoperability, and efficient management across the LSS.

Applications As Rooms in a House

To draw an analogy, consider individual applications as rooms in a house. Each room, whether it's the Kitchen, Living Room, Bathroom, Furnace Room, or Home Office, serves a specific purpose, much like how each application in a unit serves a particular function. These rooms (applications) are equipped with their own set of furniture and appliances (features and functionalities), tailored to their specific use case. They are also connected to essential resources—water, electricity, Internet, etc.—similar to how applications are connected to necessary computing resources and services. Just as a well-designed room efficiently serves its intended purpose, a well-engineered application effectively meets its functional requirements within the unit.

CHAPTER 7   IMPLEMENTATION

Figure 7-6. In Evolveburgh's exclusive Bughattan district, where each application is a fortified room, bugs with enough digital dough can rent byte-sized apartments. Our engineers only accept crypto, no wire transfers. Welcome to Bughattan—where the only bugs allowed are the ones with enough cache to crash the system in style!

Suites of Applications As Apartments

Expanding this analogy, suites of applications can be likened to apartments within an apartment building. Each apartment represents a suite, encompassing a collection of rooms (applications). These apartments are designed to offer a comprehensive living solution, just as suites of applications provide a complete set of functionalities addressing a broader aspect of the unit's operations. In this context, different rooms within an apartment collaborate seamlessly, similar to how applications within a suite integrate and interact to deliver a unified service. Often, these suites or applications are provided by vendors, offering specialized

CHAPTER 7 IMPLEMENTATION

functionalities tailored to specific needs, much like how apartments can be furnished and tailored by different builders or designers to meet diverse living requirements.

Vendor-Provided Suites and Applications

In many cases, suites or applications are sourced from external vendors. These vendor-provided solutions come with predefined functionalities and integrations, much like a furnished apartment with all necessary amenities and design elements. The choice of these suites or applications is pivotal, as it impacts the unit's capability to fulfill its purpose effectively. Integration of these vendor-provided components into the unit's architecture requires careful planning to ensure compatibility, scalability, and maintainability within the broader LSS.

The analogy of rooms and apartments provides a relatable understanding of how applications and suites of applications function within a sociotechnical unit. This perspective aids in grasping the complexity and interconnectivity of these components, highlighting the importance of thoughtful engineering and integration to achieve a harmonious and efficient unit within the LSS.

Key Takeaways

Here are the key takeaways from the chapter, highlighting the iterative and adaptive nature of implementing large sociotechnical systems. These points emphasize the importance of distinguishing between code organization and engineering, effective cloud structuring, and the analogy of rooms and apartments to understand the components of sociotechnical units.

- **Code organization vs. engineering**: By distinguishing between code organization and engineering, practitioners can close the gap between theoretical concepts and practical application, creating systems that are technologically robust, socially pertinent, and organizationally efficient.

- **Distributed composition at sociotechnical system level**: The Organization mechanism centers around the deployment of the distributed composition of the enterprise, ensuring that geographically dispersed and operationally independent units operate cohesively as a unified whole.

- **Partitioned composition at unit level**: At the unit level, the principle of partitioned composition involves organizing each unit within its designated cloud environment(s), maintaining distinct operational boundaries and capabilities.

- **Engineering mechanism**: The Engineering mechanism delves into the practical aspects of software development and integration, ensuring that applications align with the unit's defined purpose and function.

- **Effective cloud structuring**: Various strategies can be employed to structure LSS and units effectively within cloud environments, leveraging both single-cloud and multicloud configurations.

- **Naming conventions and tagging**: Developing and implementing effective naming conventions and a robust tagging strategy are vital for resource organization and management within cloud environments.

- **Multicloud approach**: Adopting a multicloud approach presents unique challenges and opportunities, requiring robust coordination, expertise in diverse platforms, and unified management tools.

- **Core building blocks of sociotechnical units**: In UOEA, engineering sociotechnical units involve a detailed focus on their core components—applications and suites of applications, ensuring consistency, interoperability, and efficient management.

- **Analogy of rooms and apartments**: Individual applications can be likened to rooms in a house, and suites of applications to apartments within an apartment building, providing a relatable understanding of their function within a sociotechnical unit.

CHAPTER 8

Interactions

The author contends that social animals may offer valuable lessons in coordination. Liberated from the confines of Business Process Management, their "Instinctive Process Management" approach emerges as a model of simplicity and effectiveness, devoid of the all-too-familiar corporate labyrinth of swimlanes and flowcharts. It seems ants and bees could run masterclasses for Fortune 500 CEOs, all without a single PowerPoint slide in sight.

In the preceding chapters, we examined the structural framework of large sociotechnical systems, which are composed of uniformly structured units. These units are classified into two principal categories: integrative units, which coordinate system-wide operations, and operational units, which execute specific tasks. Collectively, these units form the essential building blocks of LSS, each contributing distinct yet interdependent functions to the overall system.

Integrative units act as the central hubs within LSS, providing direction and facilitating coordination, decision-making, and strategic planning. They oversee the alignment of various operational units, ensuring that all activities are harmonized toward the system's overarching goals.

These units are characterized by their broad perspective, encompassing multiple operational units and focusing on long-term objectives and systemic coherence. They are pivotal in integrating diverse activities and resources across the system.

CHAPTER 8 INTERACTIONS

Operational units are the execution arms of LSS, responsible for carrying out specific tasks, processes, and operations and interacting with entities in the surrounding environment. They operate within the framework established by integrative units, focusing on the practical implementation of strategies and day-to-day activities.

These units are more specialized and purpose oriented, with a narrower focus compared to integrative units. They are designed to fulfill specific purpose(s) efficiently and effectively, contributing to the overall productivity and responsiveness of the LSS.

Each unit within LSS is a collective of human agents augmented by various technological extensions. This hybrid composition leverages the strengths of both human capabilities and advanced technologies to enhance the functionality and performance of the system.

Human agents bring essential skills, expertise, and decision-making abilities to the units. They provide the cognitive flexibility, creativity, and ethical judgment necessary for navigating complex and dynamic environments.

Human agents interact with both other humans and technological tools, forming a cohesive operational team within each unit. Their adaptability and learning capacity are essential for the continuous improvement of unit performance.

Technological extensions amplify the capabilities of human agents, enabling them to achieve higher levels of productivity and effectiveness. They facilitate seamless operations, reduce human error, and enhance overall system reliability. Robots perform physical tasks, enhance operational efficiency, and ensure precision in repetitive activities. Virtual assistants aid in information processing, communication, and routine decision-making, supporting human agents in their tasks. Tools and platforms act as the digital infrastructure that streamlines workflows, integrating disparate systems and providing a cohesive environment for data analysis and strategic planning. They serve as the connective tissue that enables diverse technological extensions to work in concert.

CHAPTER 8   INTERACTIONS

Smart objects, equipped with advanced sensors and actuators, not only gather and communicate data but also execute critical tasks autonomously, adapting to and interacting with their environment to optimize the functionality and responsiveness of the larger sociotechnical system.

By integrating human agents with advanced technological extensions, each unit within an LSS can operate synergistically, maximizing the strengths of both human and technological elements. This integration forms the backbone of large sociotechnical systems, ensuring they can adapt, scale, and respond to the ever-changing demands of their environments.

In this chapter, we shift our focus to the complex dynamics of interactions within large sociotechnical systems. Understanding these interactions is necessary for optimizing the performance and coherence of such complex systems.

We will focus on coordination as an essential mechanism that structures and gives meaning to all interactions within sociotechnical systems. Despite advancements in these areas, the industry often overlooks coordination, leading to its underdevelopment. Conventional Business Process Management systems have not adequately addressed coordination, resulting in reliance on implicit—manual and email-based—methods. These approaches are typically suboptimal and error-prone. Addressing this gap is critical for enhancing the functionality and resilience of LSS.

CHAPTER 8   INTERACTIONS

Figure 8-1. Confronting BPM limitations, Evolveburgh IT revolutionized workflow management with streamlined robotic puppetry for peak performance and robust marionette synchronization, ensuring precision in every corporate pirouette

The approach encompasses two primary dimensions of interactions: those occurring at the system level and those at the unit level.

System-level interactions refer to the exchanges and relationships that occur between different units of the LSS and between these units and entities in the surrounding environment. These interactions are fundamental to the overall functionality and adaptability of the system.

Unit-level interactions occur within individual units, focusing on the interplay between a unit's agents—both human and technological. These interactions are pivotal for the effective functioning of each unit.

By examining both system-level and unit-level interactions, this chapter provides a comprehensive understanding of the mechanisms that drive effective coordination within LSS, ultimately contributing to their success and sustainability.

System-Level Interactions

System-level interactions within large sociotechnical systems are the exchanges and relationships that occur both internally among the different units of the system and externally between these units and various entities in the environment. These interactions are pivotal in ensuring the seamless operation, adaptability, and overall effectiveness of the LSS.

In the previous chapters, the discussion centered on the nature of relations that characterize interactions within LSS. These include transactional relations that dictate exchanges between units and consumers, organizational relations that define the structure between integrative and operational units, collaborative relations that enable joint efforts among operational units, cooperative relations that underpin all types of interactions, compliance relations that maintain conformity between units and regulatory authorities, and competitive relations that manage the rivalry between units and their competitors. The nature of these relations is critical in determining the approach to coordination, influencing its design, implementation, and management within a sociotechnical ecosystem.

At the core of system-level interactions is the concept of coordination, which ensures that activities across various units are aligned with the system's objectives and responsive to external demands. As was mentioned earlier, coordination is facilitated by robust communication and connectivity mechanisms, which serve as the backbone for these interactions.

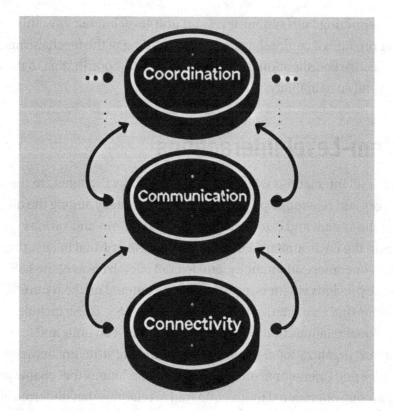

Figure 8-2. Coordination among units within a large sociotechnical system—and between these units and external entities—relies on effective communication, which is fundamentally supported by connectivity

Coordinating Transactional Interactions

Transactional interactions in large sociotechnical systems refer to the exchanges between units and external consumers or clients. These interactions involve the delivery of products, services, or information and are critical for meeting external demands, ensuring customer

CHAPTER 8   INTERACTIONS

satisfaction, and maintaining the overall reputation of the system. Effective coordination of these interactions is essential for optimizing performance and achieving strategic objectives.

To delve into the specifics, transactional interfaces come in various forms, each with distinct characteristics that cater to the coordination needs of these interactions. *Self-service interfaces* are designed to empower customers by providing them with direct access to services and information, thereby reducing the need for direct human assistance. These interfaces are characterized by their user-friendly design, intuitive navigation, and real-time feedback mechanisms, which collectively streamline the transaction process.

On the other hand, *service interfaces* are more interactive and often involve a human element. They are characterized by personalized assistance, where a representative aids the customer through the transaction. This type of interface is particularly effective in handling complex transactions or when a high degree of customization is required.

Moreover, there are automated service interfaces that combine elements of both self-service and human-assisted models. These interfaces use algorithms and artificial intelligence to guide users through a transaction, offering a balance between efficiency and personalized service.

Each type of interface requires a strategic approach to coordination, ensuring that the system can handle the volume and complexity of transactions while maintaining a high level of customer service. This involves the integration of technology, process management, and human resources to create a cohesive system that can adapt to changing demands and continuously improve the customer experience. The ultimate aim is to create a harmonious ecosystem where transactional interactions are not just exchanges but experiences that reinforce the system's value and reliability.

CHAPTER 8   INTERACTIONS

Self-Service Interfaces

Self-service interfaces are a testament to customer empowerment, providing individuals with the tools they need to independently access services and information. This empowerment is achieved through a variety of features that facilitate user engagement and autonomy. Customers can effortlessly search for products or information that align with their needs, using intuitive search functions. They can then refine their findings with filters, ensuring that the results match their specific criteria. Selection is made simple, allowing customers to choose items or services with confidence. The process of purchasing is streamlined with a virtual cart, where customers can review and manage their selections before finalizing their order. For additional support, ordering assistance and frequently asked questions are readily available, offering guidance and resolving common issues. Personalization is at the forefront, with the ability to create individual profiles and set preferences, which the system remembers for future interactions, crafting a tailored experience each time. These features collectively reduce the dependency on human assistance, making the customer's journey through the interface a smooth and efficient one.

CHAPTER 8   INTERACTIONS

Figure 8-3. *Evolveburgh's latest invention, the "Attach-O-Matic," gives new meaning to "sticking around"! With its universal grip, this self-service interface transforms anything—from your unsuspecting cat into a customer service rep to your toaster into a tech support guru. Just a heads-up: if you pause too long, you might just become the next exhibit in the "Evolveburgh Museum of Modern Art." So remember, in Evolveburgh, "letting go" isn't just emotional advice, it's a survival tip!*

Self-service interfaces are characterized by

- **Explicit processes**: The processes and options available are clearly outlined for the user, ensuring transparency and ease of use.

- **Formal structure**: They follow predefined structures and rules to ensure consistency in service delivery.

- **Direct interaction**: Users interact directly with the system without intermediaries, enhancing efficiency.

- **Machine execution**: Designed to function without human intervention, leveraging automation for consistent performance.

- **Designed interaction**: Carefully crafted to facilitate easy and intuitive user experiences.

- **Controlled environment**: Governed by algorithms and rules to maintain system integrity and reliability.

Service Interfaces

Service interfaces stand out for their interactive nature, often necessitating the involvement of a human element to accommodate the unique specifications or the presence of the customer. These interfaces are tailored to provide personalized assistance, ensuring that a representative is available to guide the customer through each step of the transaction. This human touch is especially valuable in situations where transactions are complex or require a high level of customization. By integrating personal interaction, service interfaces are able to address the specific needs and preferences of each customer, resulting in a more satisfying and effective exchange. These interfaces are characterized by

CHAPTER 8   INTERACTIONS

Figure 8-4. *Despite Evolveburgh's sophisticated service interfaces, they've yet to discover the "HU" element on their periodic table—human interaction remains their missing link*

- **Explicit/implicit procedures**: Can be explicit when procedures are defined or implicit when relying on human judgment
- **Formal/informal structure**: Formal when following strict protocols and informal when customization or discretion is involved

- **Direct/indirect interaction**: Direct when involving face-to-face or real-time communication and indirect when mediated through technology
- **Human execution**: Often rely on human expertise and decision-making
- **Designed/not designed**: Structured around designed protocols but can include elements that are not fully designed, such as personalized interactions
- **Controlled/not controlled**: Controlled within the bounds of organizational rules, but may offer flexibility in execution

Automated Service Interfaces

Automated service interfaces represent a fusion of self-service and human-assisted models, utilizing sophisticated algorithms and artificial intelligence to navigate users through transactions. These interfaces strike a harmonious balance, delivering the swift, streamlined experience of self-service while also providing the tailored, attentive interaction typically found in service interfaces. This dual approach ensures that transactions are not only efficient but also adaptively responsive to each user's unique needs and preferences. These interfaces are characterized by

- **Explicit processes**: Clearly defined processes automated for consistency.
- **Formal structure**: Adhere to strict programming and service delivery rules. Advanced AI capabilities introduce a level of informality and implicit understanding, providing responses that feel more natural and less structured.

- **Direct/indirect interaction**: Direct in providing immediate service, but can be indirect in how they interface with other system components.

- **Machine execution**: Operate through programmed algorithms and AI.

- **Designed interaction**: Engineered to combine the efficiency of machines with the nuance of human-like interaction.

- **Controlled environment**: Highly controlled to ensure reliable and predictable service delivery.

Interface Technologies

To support the diverse needs of transactional interactions, various interface technologies are employed across self-service, service, and automated service interfaces. These technologies ensure that the coordination of interactions is efficient, user-friendly, and responsive to the evolving digital landscape.

- **Web interfaces**: Common touchpoints for transactional interactions, accessible through browsers on various devices. They offer a versatile platform for delivering services and information, characterized by their broad reach and ability to provide rich, interactive experiences through multimedia content and real-time data.

- **Mobile interfaces**: Tailored for smartphones and tablets, providing a more personal and portable interaction channel. Designed with the mobile user in mind, they focus on touch interactions, location services, and streamlined processes that cater to users on the move.

- **APIs (Application Programming Interfaces):** Essential for automating processes and integrating diverse systems, enabling different software systems to communicate and exchange data seamlessly. APIs support the seamless integration of various system components, ensuring that the user's experience is consistent and responsive across different platforms and devices.

- **Conversational interfaces:** Such as chatbots and virtual assistants, offer a natural and engaging way of interaction. They mimic human conversation, allowing users to perform transactions through text or voice commands. These interfaces are increasingly sophisticated, with the ability to understand context, learn from interactions, and provide personalized responses.

- **IoT interfaces:** Represent the cutting edge of transactional interactions, where physical objects are embedded with sensors and connectivity to interact with the system. Examples include smart meters that automatically report usage data to utility providers or wearable health devices that communicate with healthcare providers.

Coordination Mechanisms

Effective coordination of transactional interactions relies heavily on robust communication and connectivity mechanisms. These mechanisms ensure that all parties involved have access to the necessary information and resources to complete transactions efficiently.

- **Integrated platforms**: Connect various units and consumers, providing a centralized system for managing transactions. They facilitate seamless data sharing, ensuring that all relevant information is available to support decision-making and process execution.

- **Real-time communication**: Tools such as instant messaging enable immediate communication between consumers and units, enhancing responsiveness and reducing delays. Automated alerts and notifications keep consumers informed about the status of their transactions, improving transparency and engagement.

Challenges and Solutions

Despite advancements in self-service and service interfaces, coordinating transactional interactions presents several challenges:

- **Complexity of transactions**: Complex transactions may require multiple steps and coordination across different units, increasing the risk of errors and delays. Implementing integrated platforms and real-time communication tools can streamline these processes, reducing complexity and improving efficiency.

- **Consumer expectations**: High consumer expectations for fast, personalized, and reliable services can be difficult to meet consistently. Enhancing self-service interfaces with advanced automation and AI capabilities can improve responsiveness and personalization, meeting consumer demands more effectively.

- **Data security and privacy**: Ensuring the security and privacy of consumer data is critical but challenging, especially with increasing regulatory requirements. Robust data security measures and compliance with regulations can protect consumer information and maintain trust.

By incorporating diverse interfaces such as web, mobile, API, conversational, and IoT, LSS can cater to a wide range of preferences and needs. Ensuring that the coordination of transactional interactions is efficient, user-friendly, and responsive to the evolving digital landscape is key. The seamless integration of these interfaces creates a cohesive user experience that aligns with the strategic objectives of the system, transforming transactional interactions into experiences that reinforce the system's value and reliability.

Coordinating Organizational Interactions

Coordinating organizational interactions within LSS involves managing the complex relationships between integrative and operational units. Integrative units provide direction and boundary conditions to ensure operational units contribute effectively to the overall success of the system. Similarly, the global integrative unit sets the framework for regional integrative units. This section explores the types of interfaces used to facilitate these interactions, focusing on directive and reporting interfaces.

CHAPTER 8   INTERACTIONS

Figure 8-5. *Evolveburgh's scientists finally cracked the code on sociotechnical systems: it turns out, the integrative components and operational components have been in a complicated relationship this whole time. They're like the hardware and software of a dating app—totally dependent on each other, but still can't agree on anything. Love-hate relationships have never been more technical!*

Directive Interfaces

Directive interfaces are mechanisms through which integrative units set processes, rules, and expectations for operational units. These interfaces can be structured (formal business rules) or unstructured (documents and informal guidelines).

Structured Business Rules Interfaces

- **Budget constraints**: This interface provides a clear framework for financial limitations. It communicates available resources and allocates them across various operational units, ensuring that spending aligns with strategic priorities. Budget constraints help operational units plan and execute their activities within the set financial limits.

309

- **Performance expectations**: This interface sets the expected outcomes and efficiency metrics that operational units must meet. It includes KPI targets, efficiency metrics, and outcome benchmarks, driving continuous improvement and aligning operational efforts with strategic goals.

Unstructured Documents and Guidelines

- **Ad hoc meetings**: Video conferencing tools like Zoom or Microsoft Teams enable quick setup of virtual meetings, allowing real-time collaboration from different locations. These tools support spontaneous discussions and decision-making processes.

- **Rules of conduct**: Digital handbooks or intranet portals host the organization's rules of conduct, making them easily accessible to all employees. These resources ensure that everyone is aware of and adheres to organizational standards.

- **Enterprise standards**: Document management systems like SharePoint or Confluence store and disseminate quality, operational, and safety standards. These systems ensure that all units have access to the necessary guidelines and protocols.

- **Cultural norms**: Digital newsletters and webinars promote cultural initiatives, reinforcing organizational values and norms through interactive online events.

CHAPTER 8   INTERACTIONS

- **Communication protocols**: Email systems and communication platforms configured to follow specific protocols ensure that the right information reaches the appropriate parties, facilitating structured and efficient communication across units.

Reporting Interfaces

Reporting interfaces facilitate the flow of information from operational units back to integrative units, providing essential feedback for decision-making and strategic adjustments. These interfaces are typically structured to ensure that data is conveyed in a consistent and measurable manner.

- **Performance dashboards**: Visual tools that display key performance indicators (KPIs) to track progress against goals. These dashboards provide real-time data and insights into operational performance, enabling integrative units to monitor and guide activities effectively.

- **Progress reports**: Regularly submitted documents that outline achievements, challenges, and future plans. Progress reports provide a comprehensive view of operational activities, helping integrative units understand the current state and plan future actions.

- **Incident logs**: Detailed records of events that deviate from the norm, providing insights into areas that may require attention. Incident logs help identify issues, assess their impact, and implement corrective measures to prevent recurrence.

311

CHAPTER 8 INTERACTIONS

Leveraging Technology for Effective Coordination

By leveraging both directive and reporting interfaces, organizations can ensure that necessary information and standards are not only accessible but also integrated into the daily workflow. This integration promotes efficiency, compliance, and continuous improvement across the LSS. The key to successful coordination lies in the seamless integration of these interfaces, creating a cohesive operational environment that aligns with the strategic objectives of the system.

Through this multifaceted approach, LSS can effectively manage the dynamic interactions between integrative and operational units, fostering a collaborative and responsive organizational culture.

Coordinating Collaborative Interactions

Coordinating collaborative interactions within large sociotechnical system (LSS) involves managing the inter-unit operational activities and strategic initiatives. This coordination ensures that all units work synergistically toward the common goals of the organization. In this section, we explore the mechanisms for coordinating these interactions, focusing on asynchronous communication, human oversight, and the potential of stigmergy.

CHAPTER 8   INTERACTIONS

Figure 8-6. *Evolveburgh's latest invention, recursive collaboration, turns teamwork into an endless loop! Meeting marathons stretch on forever, brainstorming picnics are all-you-can-think buffets on permanent Groundhog Day, and elevator pitches transport you through infinite realities. Our mascot, Deja Vu the hamster, runs tirelessly on his wheel, symbolizing this never-ending cycle. Those who tried to escape by inventing a time machine discovered it's just another meeting room for timeless talks where every hour is a discussion about the space-time continuum. The snacks are great, though—they're always from the future, so they never expire, and the coffee? It's so déjà brew, it's worth repeating the same day for!*

Indirect Coordination Through Asynchronous Communication

Indirect coordination is pivotal in managing collaborative interactions across different units within an LSS. This approach leverages a combination of pub/sub notifications and API interfaces to enable indirect communication and access to information.

CHAPTER 8   INTERACTIONS

- **Notifications**: Pub/sub systems and broadcasting methods allow units to be notified of changes without the need for direct, synchronous communication. This method ensures that relevant updates are disseminated efficiently across the organization.

- **API interfaces**: Notifications provide awareness, while API interfaces offer ad hoc access to detailed information as needed. This on-demand access allows units to delve deeper into the particulars of a notification when it is pertinent to their operations.

Human Coordination and Direct Mechanisms

Despite the advancements in coordination technologies, the maturity level of these systems is still evolving, and the implementation of Business Process Management (BPM) has not always met success. This is primarily due to the complex and nuanced nature of organizational dynamics that technology alone cannot fully capture. Human coordination remains essential, as it provides critical oversight and decision-making capabilities that automated systems have yet to replicate effectively. Humans possess the unique ability to interpret the context of notifications, assess the relevance of data, and steer the strategic direction of the organization. Their judgment and insight are invaluable, particularly when technology falls short in understanding the subtleties of human interactions and the intricate web of organizational processes.

- **Human oversight**: Human coordinators interpret the context of notifications, assess the significance of data, and guide the overall strategic direction. This human element provides oversight and decision-making capabilities that automated systems cannot fully replicate.

- **Business Process Management**: BPM provides a structured method for overseeing and aligning tasks with the organization's strategic objectives. While effective in some scenarios, the current maturity of BPM often limits its viability, necessitating more adaptive coordination mechanisms.

Potential of Stigmergy

Stigmergy, an indirect coordination mechanism inspired by social insects, offers promising potential for discovering and formalizing effective coordination patterns within LSS.

- **Environmental cues**: Similar to how insects leave traces to signal others, digital systems can be designed to leave behind "cues" that inform subsequent actions of other units. These cues facilitate indirect coordination through the environment itself.

- **Emergent patterns**: Over time, these environmental cues can lead to emergent patterns of behavior that are self-organizing and adaptive to changes within the system. This enhances the overall responsiveness and efficiency of collaborative interactions.

- **Machine learning analysis**: Applying machine learning techniques to analyze and optimize these emergent patterns can lead to more efficient pathways for coordination and decision-making. This leverages the collective intelligence inherent in the system.

CHAPTER 8   INTERACTIONS

Embracing Stigmergy

Embracing stigmergy represents a paradigm shift toward a more decentralized and autonomous approach to organizational coordination. This approach leverages the collective intelligence of the system, fostering a dynamic environment where collaboration is driven by both direct interactions and the shared digital ecosystem.

- **Paradigm shift (decentralization and autonomy):** Stigmergy encourages a move away from centralized control toward decentralized, self-organizing coordination. This shift enhances adaptability and resilience within the system.

- **Practical implementation (observation and codification):** By observing interactions and adjustments made in response to asynchronous notifications and API data, organizations can identify and codify best practices that emerge organically. This process leads to the formalization of effective coordination patterns based on real-world dynamics.

Looking Forward

The concept of stigmergy presents an opportunity to further refine coordination mechanisms within LSS. Analogous to the way pedestrians create desire paths as the most efficient routes across a landscape, organizations can observe and codify stigmergic coordination patterns to establish the most effective pathways for collaboration. By harmonizing stigmergic principles with asynchronous communication methods, organizations can develop a robust coordination framework. This framework not only manages complex collaborative activities but also learns and adapts over time. As these "desire paths" of coordination are formalized, the system can evolve into an adaptive and self-optimizing entity, one that smartly integrates advanced technological systems with the nuanced judgment and oversight of human cognition.

To highlight stigmergic coordination's emergent, adaptive, and user-driven nature, it's apt to draw an analogy to desire paths. Desire paths are naturally formed trails created by repeated human use, showcasing the most efficient and practical routes based on real-world behavior rather than predetermined planning. Similarly, stigmergic coordination evolves through the continuous interactions and adaptations of the system's users, creating an efficient and flexible network that responds to actual needs. Both concepts emphasize organic, user-driven patterns that emerge through repeated individual actions rather than predesigned structures. Here are some parallels to support this analogy:

- **Natural formation**: Just as desire paths form through repeated use and adaptation to the environment, stigmergic structures evolve as agents respond to environmental cues, creating informal "paths" of coordination without a central plan.

- **Responsiveness to needs**: Desire paths emerge from people taking the most efficient or comfortable routes, often revealing needs or preferences not anticipated in the original design. Similarly, stigmergic coordination is responsive, as each agent interacts with cues in a way that best serves immediate goals, leading to adaptive, collective solutions.

- **Self-optimization**: Both desire paths and stigmergy optimize for efficiency over time. In stigmergy, as agents interact repeatedly in an environment, their behaviors reinforce certain "paths" or patterns that streamline coordination, akin to how a frequently used desire path becomes more defined, creating a natural shortcut.

- **Lack of explicit structure**: Like desire paths, which lack formal boundaries or materials and rely on the accumulated "traces" of others' footsteps, stigmergic coordination structures lack explicit, prescriptive rules. Instead, they rely on environmental cues, indirect signals, and the collective knowledge embedded in the actions of others.

- **Adaptability and flexibility**: Desire paths can change with shifting user needs or seasonal conditions, remaining flexible and responsive to context. Similarly, stigmergic coordination adapts dynamically, responding to shifts in environmental inputs and the behaviors of other agents.

In summary, the coordination of collaborative interactions in a sociotechnical system is a delicate balance between technology and human oversight. Asynchronous communication enabled by pub/sub systems and complemented by the use of API interfaces provides a strong foundation, while the integration of stigmergic principles offers a pathway to a more adaptive and self-organizing future. This combination forms a robust framework for managing the complex web of organizational activities, enhancing efficiency, responsiveness, and innovation within the system.

Coordinating Compliance Interactions

In the context of LSS, coordinating compliance interactions involves a twofold process. Firstly, it encompasses the interactions of internal units, such as compliance and legal units, with external entities like authorities and regulators. These units act as the primary interface between the organization and external regulatory bodies, ensuring that the organization's operations align with the relevant laws and regulations.

They engage in ongoing dialogue with authorities, interpret regulatory requirements, and translate them into actionable guidelines for the organization.

Figure 8-7. *In Evolveburgh, the bond between businesses and regulators is so harmonious that paying taxes is practically a spiritual experience—like adding good vibes to the city's karma bank*

The second aspect of coordinating compliance interactions involves the dissemination of the interpreted information and its implementation across other units within the LSS. Once the compliance and legal units have interpreted the regulatory requirements, they communicate this information to the rest of the organization. This could involve creating comprehensive compliance manuals, conducting training sessions, or using digital platforms to share updates.

The aim is to ensure that all units within the LSS are aware of the compliance requirements and understand how to implement them in their day-to-day operations. This dual approach to coordinating compliance interactions ensures that the organization not only meets external regulatory requirements but also fosters a culture of compliance internally, enhancing overall operational efficiency and integrity.

CHAPTER 8   INTERACTIONS

Coordinating Competitive Interactions

Coordinating competitive interactions within LSS involves a strategic approach to understanding, responding to, and engaging with competitors in the surrounding environment. These interactions are vital for maintaining a competitive edge and driving innovation within the organization.

Figure 8-8. Evolveburgh's "Strategy Arena" hosts epic battles of wits using Porter's five forces, all to keep that "Competitive Edge." Modern business meets medieval showdown—because spreadsheets are so last century

Competitive interactions begin with the systematic monitoring and analysis of competitors' activities, strategies, and performance. This could involve tracking market trends, technological advancements, and shifts in consumer behavior that might impact the competitive landscape. Units within the LSS, particularly those involved in strategy and market intelligence, play a key role in this process.

Once this information is gathered, it's important to disseminate it effectively within the LSS. This ensures that all units are aware of the competitive context and can align their actions accordingly. For instance, the product development unit might adjust its innovation strategy based on competitor activities, while the marketing unit might refine its messaging to highlight the organization's unique value proposition.

Moreover, competitive interactions also involve proactive engagement with the competitive environment. This could include strategic initiatives aimed at outperforming competitors, such as launching new products, entering new markets, or forming strategic partnerships.

In addition to these traditional competitive strategies, there's also room for constructive interactions with competitors, often referred to as "coopetition." This involves cooperating with competitors in areas where mutual benefits can be derived while still maintaining competition in other areas. Coopetition can lead to shared benefits like cost savings, access to new markets, increased innovation, and the exchange of knowledge and skills.

In essence, coordinating competitive interactions in LSS involves a dynamic and ongoing process of understanding, responding to, and proactively engaging with competitors. This not only helps the organization maintain its competitive position but also fosters a culture of continuous learning, adaptation, and constructive collaboration.

Coordinating Cooperative Interactions

Cooperative interactions within LSS enhance and complement all other types of interactions, including transactional, organizational, collaborative, compliance, and even competitive interactions. Effective coordination of cooperative interactions ensures that units within the organization can work together harmoniously, sharing information and resources to achieve common goals.

CHAPTER 8   INTERACTIONS

Figure 8-9. During an intense debate, Evolveburgh's mayor tried to explain the concept of cooperation by referencing the "prisoner's dilemma." But as he dove into the details using slang that would make a mob boss blush, he almost spilled the beans. The council members were so perplexed that even the town's notorious cat burglar halted his escapade to consult a dictionary!

While essential coordination efforts supply the necessary information and data for other types of interactions, cooperation goes a step further. It enhances strategic alignment and flexibility, adding a layer of depth and effectiveness. This is achieved through the facilitation of shared expertise and knowledge, which goes beyond what is required, fostering innovation and collective problem-solving.

Cooperation also aids in the reduction of redundancies by sharing additional information, which helps identify and eliminate overlaps, thereby streamlining processes and improving overall efficiency. Units can share strategic insights and long-term plans through cooperation, aligning their efforts more closely with the organization's overarching goals. Furthermore, cooperation allows units to adapt their strategies based on shared experiences and insights, enhancing flexibility and responsiveness to changes. This exploration of the importance of cooperation highlights what can be shared beyond the essential information in various types of interactions.

What Can Be Shared in Cooperative Interactions

Beyond the necessary information, cooperative interactions can involve sharing the following additional elements:

- **Market insights**: Sharing insights about market trends and customer feedback can help units refine their offerings and improve customer satisfaction.

- **Best practices**: Units can share best practices in transaction handling, improving efficiency and service quality.

- **Strategic plans**: Sharing detailed strategic plans and objectives helps align the efforts of various units with the organization's long-term goals.

- **Project updates**: Regularly sharing updates on collaborative projects keeps all involved units informed and engaged, facilitating smoother collaboration.

- **Technical expertise**: Units can offer technical expertise and support to each other, enhancing the overall capabilities of the organization.

- **Regulatory updates**: Sharing updates on regulatory changes and compliance requirements ensures all units are aware and can adapt accordingly.

- **Audit results**: Units can share findings from internal audits to improve compliance and address any identified issues proactively.

- **Competitive analysis**: Sharing insights from competitive analysis helps units understand market positioning and develop more effective strategies.

CHAPTER 8   INTERACTIONS

- **Innovation strategies**: Units can collaborate on innovation initiatives, leveraging shared knowledge to stay ahead of competitors.

Mechanisms for Cooperative Interactions

The coordination of cooperative interactions is practically achieved through a blend of communication methods and resources. Asynchronous communication methods, such as publish/subscribe (pub/sub) messaging systems and emails, play a significant role in this process. These methods allow units to send and receive updates at their own pace, reducing the need for simultaneous presence and thus enhancing flexibility.

Asynchronous communication, facilitated by pub/sub systems and emails, allows for notifications about updates, changes, or new insights to be disseminated effectively. This ensures that all units stay informed without the need for constant real-time interaction, allowing them to manage their tasks more flexibly and efficiently.

CHAPTER 8   INTERACTIONS

Figure 8-10. *"In Evolveburgh, asynchronous communication became the latest fashion trend, with locals rocking time-delayed replies like it's the hottest accessory. Now, everyone's out of sync—but looking fantastic doing it. Local brand Pubsub's "Why reply now?" is outpacing Nike's "Just Do It" because who needs instant when you can have style?!"*

While asynchronous notifications provide high-level updates, detailed information exchanges are often necessary for specific tasks. These more direct interactions are facilitated through static resources, APIs, and other data sources. Units make direct calls to these resources to retrieve detailed information relevant to their tasks, ensuring that specific, actionable data is available on-demand and complementing the broader updates received through asynchronous notifications.

The combination of asynchronous notifications and direct information retrieval enables a comprehensive and efficient coordination of cooperative interactions. This blend ensures that units receive broad updates and high-level notifications that keep them aligned with the organization's goals and aware of relevant changes across the system.

CHAPTER 8   INTERACTIONS

Additionally, units can access detailed information necessary for executing their specific tasks effectively, ensuring they have all the resources they need to perform their roles without missing critical details.

Achieving a balance between asynchronous notifications and direct information retrieval is key to the successful coordination of cooperative interactions within an organization. This approach allows units to remain flexible, well-informed, and capable of accessing detailed information as needed, fostering a cooperative environment that enhances overall efficiency and effectiveness.

Unit-Level Interactions

Unit-level interactions occur within individual units through the interplay between human agents, artificial agents, and other technological extensions. These interactions are pivotal for the effective functioning of each unit, as they integrate the unique capabilities of humans and technology to achieve operational goals. By optimizing the collaboration between these elements, units can enhance productivity, innovation, and adaptability within the sociotechnical system.

Direct Coordination

Direct coordination within unit-level interactions often relies on BPM systems. BPM provides a structured and systematic approach to managing and optimizing the various processes that occur within a unit. This method focuses on predefined workflows, clear protocols, and explicit communication channels to ensure that tasks are executed efficiently and consistently.

Key Attributes of Direct Coordination

BPM-inspired coordination is characterized by structured workflows, explicit control mechanisms, well-defined roles, and a continuous cycle of monitoring and optimization. The salient features include

- **Streamlined processes**: BPM systems leverage predefined workflows to orchestrate task execution. These workflows, crafted in alignment with best practices and organizational standards, foster operational consistency and efficiency.

- **Logical task sequencing**: Tasks are arranged in a logical sequence, delineating clear dependencies and milestones. This systematic structure mitigates bottlenecks and ensures the completion of each step before transitioning to the subsequent one.

- **Unambiguous communication protocols**: BPM systems institute clear communication protocols, specifying the information to be communicated, the recipients, and the timing. This clarity reduces potential misunderstandings and ensures collective alignment.

- **Uniform documentation**: All communications and documentations adhere to standardized formats, facilitating progress tracking, issue identification, and record maintenance.

- **Explicit role definition and accountability**: BPM systems distinctly define the roles and responsibilities of each participant, both human and artificial, involved in the process. This clarity aids in appropriate task assignment and fosters accountability.

- **Role-based access control**: Access to information and resources is regulated based on roles, ensuring that individuals possess the necessary permissions to execute their tasks while upholding security protocols.

These features collectively enhance productivity, innovation, and adaptability within the larger sociotechnical system.

Strengths and Limitations of Direct Coordination

- **Consistency**: BPM ensures that processes are carried out consistently, reducing variability and errors.
- **Efficiency**: By following predefined workflows, tasks are completed efficiently, with minimal delays.
- **Accountability**: Clear role definitions and monitoring mechanisms enhance accountability and performance tracking.
- **Transparency**: Standardized documentation and real-time monitoring provide transparency into process execution and outcomes.

Limitations of Direct Coordination

- **Rigidity**: BPM systems can be rigid, making it difficult to adapt to unexpected changes or unique situations.
- **Complexity**: Designing and maintaining BPM systems can be complex and resource-intensive.
- **Dependence on accurate design**: The effectiveness of BPM depends on the accuracy of the predefined workflows and protocols. Poorly designed processes can lead to inefficiencies and bottlenecks.

- **Effective process mining**: BPM systems require effective process mining to analyze and improve existing processes. Ineffective process mining can lead to suboptimal results, regardless of the accuracy of the initial design.

Direct coordination through BPM systems plays an important role in unit-level interactions within LSS. By providing structured workflows, explicit communication protocols, and real-time monitoring, BPM ensures that tasks are executed efficiently and consistently. Despite its limitations, BPM is a powerful tool for optimizing processes and enhancing the collaboration between human agents and technological extensions, contributing to the overall effectiveness and adaptability of the unit.

Indirect Coordination

Indirect coordination mechanisms, inspired by swarm intelligence and stigmergy, offer a viable alternative to direct coordination in unit-level interactions within LSS. These mechanisms provide an effective means for discovering and formalizing coordination patterns, similar to the way desire paths emerge organically based on actual usage. One advanced approach within this framework is Directed Evolution, which significantly accelerates the discovery and formalization of processes by leveraging both AI and human judgment.

Stigmergy and Directed Evolution

Stigmergy is a form of indirect coordination where agents interact through modifications to their environment rather than direct communication. In stigmergy-inspired systems, agents leave traces or cues in the environment, such as digital markers, logs, or updates to shared data repositories. These traces, which can be generalized as events (messages plus timestamps), provide contextual information and status updates that influence the subsequent actions of other agents. For example, an agent might update a shared database with the status of a task, leaving a cue for other agents to either continue the task or adjust their actions accordingly. This approach harnesses the collective intelligence of the system, allowing for adaptive and self-organizing behaviors.

Directed Evolution involves the continuous monitoring of information flows, learning from emerging patterns of behavior, and making intentional modifications to guide future behavior. This method uses AI to analyze patterns and human judgment to refine and adjust the rules by which agents interact. Directed Evolution accelerates the process of identifying key events and optimizing interaction rules.

Stigmergic vs. BPM-Executed Coordination

Stigmergic coordination is driven by events, with agents reacting to environmental changes caused by other agents' actions. On the other hand, BPM coordination is task oriented, with agents executing tasks in a specific sequence as per predefined workflows and protocols.

Demonstrating high flexibility, stigmergic coordination allows agents to adjust their behavior based on real-time environmental cues. In contrast, BPM coordination, being relatively rigid, requires agents to adhere to strict workflows, which may not easily adapt to unexpected changes or unique situations.

Patterns in stigmergic coordination emerge organically from the interactions between agents and their environment, with Directed Evolution playing a key role in formalizing these patterns. Conversely, in BPM-executed coordination, the coordination patterns can be a mix of predesigned workflows and patterns discovered through process mining.

The table below outlines the structural and functional differences between direct and indirect coordination. It demonstrates how BPM-based coordination is more prescriptive and centralized, whereas stigmergy-ased coordination utilizes environmental cues, fostering a more flexible and adaptive system.

Table 8-1. Comparative characteristics of direct and indirect coordination

Criterion	Direct Coordination (BPM Based)	Indirect Coordination (Stigmergy Based)
Design	Top-down, predefined workflows	Bottom-up, emergent patterns
Communication	Explicit, often formal, direct communication	Implicit, information embedded in environment
Awareness	High situational awareness through explicit instructions	Low direct awareness, agents react to environmental cues
Adaptability	Limited by rigid processes, slower to adapt to changes	Highly adaptable as agents respond dynamically to changing stimuli
Monitoring and Control	Centralized control and tracking; progress monitored against predefined steps	Decentralized, self-regulating based on feedback from the environment
Emergence	Limited; behaviors and outcomes are predetermined by the process	High; complex behaviors and solutions emerge from individual actions
Resource Allocation	Explicitly assigned through BPM systems	Implicit, resources are accessed and utilized as needed
Roles	Clearly defined roles and responsibilities	Roles are fluid and often emergent based on situational need
Coordination Basis	Task based, each step aligned to a role and responsibility	Event based, each agent responds to environmental triggers

(*continued*)

CHAPTER 8　INTERACTIONS

Table 8-1. (*continued*)

Criterion	Direct Coordination (BPM Based)	Indirect Coordination (Stigmergy Based)
Coordination Mechanism	Process-oriented, structured workflow	Worknet oriented, unstructured, and decentralized
Coordination Engine	BPM engine that manages workflow steps and sequences	Environment or platform where agents interact and leave signals
Trigger	Tasks trigger subsequent steps; explicit hand-offs and dependencies	Environmental changes and signals trigger responses from agents

Strengths of Stigmergic Coordination

- **Adaptability**: Highly adaptive to changes and unexpected events, as agents can dynamically respond to new environmental cues.

- **Scalability**: Easily scalable, as the system can accommodate a large number of agents interacting through simple rules.

- **Emergent behavior**: Facilitates the emergence of complex behaviors from simple interactions, leveraging collective intelligence.

- **Autonomy**: Agents are empowered with sufficient autonomy, reducing the need for managers to oversee every task. This decentralization allows for more efficient and self-sufficient operation at the unit level.

- **Continuous improvement**: Directed Evolution allows for continuous monitoring and improvement of coordination patterns based on real-time data and AI analysis.

Limitations of Stigmergic Coordination

- **Complexity**: Every agent subscribed to an event may decide to execute a task or a set of tasks based on its interpretation of the event. This can result in a high degree of variability in how tasks are executed, making it difficult to predict and manage outcomes. This inherent complexity can lead to difficulties in ensuring consistency and reliability. Managers may find it challenging to oversee and control processes, as the behavior of the system can be less predictable than with BPM. This unpredictability can also make it harder to troubleshoot issues and implement standardized procedures across the system. The challenge lies in identifying and capturing only the key events, ensuring that minute events do not pollute process definitions. Effective visualization and filtering of events are essential to maintaining clarity and focus on significant actions.

- **Overhead**: Continuous monitoring and analysis required for Directed Evolution can introduce additional overhead and complexity.

- **Initial learning curve**: Implementing and fine-tuning stigmergy-inspired systems may require significant initial effort and learning.

CHAPTER 8   INTERACTIONS

Stigmergy-inspired indirect coordination, coupled with Directed Evolution, offers a flexible and adaptive alternative to traditional BPM-executed coordination. By allowing patterns to emerge organically and using AI to continuously refine these patterns, organizations can achieve a high degree of adaptability and efficiency in their unit-level interactions. While this approach comes with its own set of challenges, the benefits of emergent behavior, scalability, and continuous improvement make it a powerful tool for managing complex and dynamic environments within LSS.

Fundamental Principles of LSS Interactions

The interactions within LSS are governed by a set of fundamental principles that ensure effective coordination, communication, and collaboration among the diverse units and agents involved. These principles provide a framework for understanding how human agents, artificial agents, and technological extensions can work together harmoniously to achieve organizational goals. By adhering to these guiding principles, LSS can optimize their operations, adapt to changing environments, and maintain high levels of efficiency and innovation.

Principle 1: Coordination As the Essence of Interactions

What: UOA posits that all interactions, whether within LSS units or between LSS units and the external environment, are fundamentally about coordination.

Why: Coordination is the central thread that weaves individual unit activities into a cohesive whole. It ensures that units work toward shared goals without conflict and with an understanding of their role within the larger system. By framing interactions as coordination, UOA promotes a structured yet dynamic interplay between different components of the LSS.

Principle 2: Multifaceted Relationships Through Coordination

What: Interactions within units and with external entities, as well as their coordination and communication patterns, should be clearly differentiated and managed according to the nature of the relations between units (and between units and external entities). These relations can be exchange, organizational, collaborative, cooperative, competitive, and compliance based, among others, each necessitating a unique approach.

Why: The adoption of a one-size-fits-all strategy in understanding, managing, and evolving interactions often leads to overly complex and entangled communication, thereby increasing coordination and communication needs. It is essential to differentiate interactions based on relationship types. By aligning approaches with the specific nature of each relation type, the system can manage interactions more effectively and appropriately. This ensures that communication and coordination are context specific and optimally efficient within the sociotechnical framework.

Principle 3: Cooperative Interactions to Enhance Relationships and Innovation

What: Cooperative interactions go beyond the required minimum in other types of interactions to strengthen relationships, foster innovation, and improve overall system performance.

Why: By engaging in cooperative interactions, units can share additional insights, resources, and support that are not strictly necessary but highly beneficial. This approach enhances the effectiveness of transactional, organizational, collaborative, compliance, and competitive interactions. Cooperative efforts help build stronger relationships, encourage knowledge sharing, and foster a culture of innovation. This collaborative environment not only improves immediate task outcomes but also contributes to long-term organizational growth and resilience. By prioritizing cooperation, LSS can create a more dynamic, supportive, and innovative ecosystem.

Principle 4: Structured Communication Protocols

What: Communication between units should be standardized, with message headers clearly defining the nature of the relationship and relevant coordination information.

Why: Standardized communication protocols ensure that messages are interpreted correctly and acted upon efficiently, reducing the chances of miscommunication. Clear headers help in routing messages appropriately and facilitate automated processing, which is key for maintaining the tempo of operations within an LSS.

Principle 5: Indirect Coordination and Communication

What: UOA emphasizes the use of indirect coordination mechanisms, such as stigmergic coordination, and indirect communication patterns like publish/subscribe (pub/sub).

Why: Indirect coordination and communication reduce the dependencies between units, enabling them to act autonomously while still contributing to the system's objectives. This increases the system's resilience, as it does not rely on a central point of control that could become a bottleneck or a single point of failure. Moreover, it aligns with the distributed nature of LSS, allowing for greater scalability and flexibility.

CHAPTER 8   INTERACTIONS

Principle 6: Asynchronous Communication and Direct Information Access

What: The coordination of most inter-unit interactions is primarily achieved through asynchronous communication methods, complemented by direct API calls for detailed information based on confidentiality needs.

Why: Asynchronous communication methods, such as publish/subscribe messaging systems and emails, allow units to send and receive updates at their own pace, reducing the need for simultaneous presence and enhancing operational flexibility. This approach ensures that all units stay informed of relevant changes and updates without constant real-time interaction, which can be disruptive and inefficient. Direct API calls provide a mechanism for accessing detailed and specific information on demand, ensuring that sensitive data is securely shared only with authorized units. This combination of asynchronous notifications and direct information retrieval supports a balanced and efficient coordination framework, enhancing the overall scalability, responsiveness, and confidentiality of the LSS.

CHAPTER 8   INTERACTIONS

Key Takeaways

This chapter has highlighted the pivotal role of coordination within Unit Oriented Enterprise Architecture in structuring and giving meaning to interactions within large sociotechnical systems. By understanding the dual dimensions of system-level and unit-level interactions and the mechanisms that facilitate these processes, we can optimize the efficiency, adaptability, and resilience of LSS.

- **Coordination as the cornerstone**: UOA focuses on coordination as an essential mechanism that structures and gives meaning to all interactions within sociotechnical systems.

- **Dual dimensions of interactions**: The approach encompasses two primary dimensions of interactions—those occurring at the system level and those at the unit level.

- **System-level interactions**: These refer to the exchanges and relationships between different units of the LSS and between these units and entities in the surrounding environment, fundamental to the overall functionality and adaptability of the system.

- **Unit-level interactions**: Occurring within individual units, these focus on the interplay between human and technological agents, pivotal for the effective functioning of each unit.

- **Transactional interactions**: Predominantly coordinated using self-service interfaces, service interfaces, and automated service interfaces.

- **Asynchronous communication and direct API calls**: Organizational, collaborative, and cooperative interfaces between LSS units are primarily coordinated through asynchronous communication methods, complemented by direct API calls for detailed information.

- **Direct and indirect coordination mechanisms**: Unit-level interactions are managed using both direct coordination (relying on BPM systems) and indirect coordination (inspired by swarm intelligence and stigmergy). Indirect mechanisms offer an effective means for discovering and formalizing coordination patterns, similar to the organic emergence of desire paths.

CHAPTER 9

Conclusion

As we conclude this comprehensive exploration of large sociotechnical systems, it's time to pause and reflect. We've traversed complex theories, avoided potential pitfalls, and navigated through intricate designs. Like any experienced professional, we understand that the realm of LSS is dynamic and unpredictable. As we conclude this phase, remember: the process of discovery is ongoing. Each new development in innovation presents fresh challenges and, with them, chances to embark on new endeavors. Until we meet again, may your endeavors in LSS be continually successful and your knowledge consistently grow.

The growing need for a more effective enterprise architecture style is becoming increasingly evident. Critiques from foundational figures like Jay Forrester ("Organizations built by committee, by intuition, and by historical happenstance—often work no better than would an airplane built by the same methods") and more recently by John Zachman ("The enterprises of today have never been engineered") are acknowledged, yet more pressing issues—such as cloud migration, building next-generation data platforms, enhancing cybersecurity, modernizing legacy systems, and implementing DevOps practices—often overshadow the problems related to enterprise architecture, especially in the absence of a holistic architectural vision.

CHAPTER 9 CONCLUSION

Despite many strategic and tactical advancements, today's frameworks often focus on aligning IT and business strategy rather than providing a clear answer to what the enterprise as a whole looks like—its components and their linkages. This gap leaves organizations "incrementally building" without a comprehensive structural design that connects the enterprise's building blocks, a shortcoming that must be addressed by clearly defining these components and their relationships. In the words of David Heinemeier Hansson, we need to "liberate the good ideas from the bullshit. To flush out all the nonsense, and to compress the remaining intrinsic complexity, so mere mortals without an advanced degree in enterprisey gibberish could employ the hidden gems of ingenuity that really were present." Although Hansson used these words to describe the creation of Ruby on Rails from lessons learned with Java Enterprise Edition, they are equally applicable to enterprise architecture, which demands a lightweight, universally understood approach to designing the modern enterprise.

Moreover, the existential risks introduced by AI underscore the need for a clear architectural approach that keeps humans in control of AI developments, thereby preserving a favorable trajectory for human evolution. Defining architectural components that comprise a social core extended by a technological periphery—not the other way around—establishes essential constraints that balance accountability with empowerment, guiding the ethical and safe development of AI systems. This approach ensures that technology enhances human capabilities rather than diminishes them. A human-centric architectural framework is critical for mitigating potential misuse and fostering the responsible, transparent evolution of AI. As we continue to push the boundaries of innovation, we must keep in mind that our technological advancements should serve humanity, fostering progress and improving quality of life, while maintaining robust safeguards to prevent unintended consequences.

This situation highlights the urgency of finding innovative solutions and underscores the importance of rethinking our architectural strategies.

Rearchitect to Mitigate Existential Risks

Rather than viewing the rapid evolution of AI solely as a source of concern, it should also be seen as a catalyst for developing a more comprehensive EA style. This architectural shift is not just a reactive measure to potential risks; it is a proactive strategy aimed at leveraging the transformative power of AI. This approach complements other measures to mitigate AI-related risks, such as establishing international regulations, fostering global collaboration, developing risk management strategies, conducting focused research, and devising strategies to counteract biases within AI systems. These actions should not be seen as isolated activities, but as integral components of a holistic approach to ensuring AI safety and security.

Firstly, the need to rearchitect the world in a way that sustains the evolutionary trajectory favoring human civilization's advancement is paramount. Over centuries, we have transitioned from passive objects to active subjects in the evolutionary process, which enabled us to consciously shape our evolutionary path. The convergence of three forms of evolution—biological evolution, evolution by organization, and evolution by extension—signals a new paradigm where humans are not merely products of evolution but active participants shaping their destiny. To maintain this evolutionary path, we must prioritize and formalize the architecture of "the social core enhanced by technological extensions" over "the technological core and social extensions," making it explicit, visible, manageable, and enforceable.

Secondly, the question of what specifically to architect arises. From our perspective, the world comprises sociotechnical systems. Analogous to a galaxy of constellations, each with its unique stars and planets forming sophisticated patterns, we can simplify our understanding of the world by viewing it as a collection of sociotechnical systems. These systems encapsulate the complex interplay between social and technological

CHAPTER 9   CONCLUSION

components, mirroring a galaxy's interconnected celestial bodies. At the heart of these sociotechnical constellations are fundamental activities: the manipulation of matter (processing, transporting, and storing); the manipulation of representations, as seen in computing where data and information undergo similar transformations; and beyond the physical and informational, the essence of human interaction forms another cornerstone, characterized by a spectrum of relational dynamics such as association, coordination, collaboration, and cooperation. These dimensions collectively weave a complex mosaic, illustrating how the confluence of human endeavors and technological advancements constructs the multifaceted sociotechnical systems that underpin and propel the world's activities forward.

Thirdly, Unit Oriented Enterprise Architecture and Purpose Driven Design provide answers to how to architect these sociotechnical systems and their building blocks—sociotechnical units, the uniformly structured building blocks of sociotechnical systems. Sociotechnical units—the units of survival and evolution—offer a comprehensive and flexible framework for understanding and designing AI systems. This approach ensures that AI agents are not seen in isolation but as part of a larger system that includes human and other technological elements.

Thus, the book *Unit Oriented Enterprise Architecture: Constructing Large Sociotechnical Systems in the Age of AI* introduces a pragmatic and practical architectural approach. This approach not only simplifies the understanding and communication of large systems by making them visible and subjecting them to continuous improvement but also harmonizes social, digital, and physical structures, enhances accountability, and optimizes communication flows.

CHAPTER 9   CONCLUSION

What's Next

As we conclude this exploration of UOEA and its transformative potential, it is vital to consider the future directions and implications of adopting this approach. The adoption of UOEA represents not merely a new architectural framework but a strategic pivot toward a more resilient, adaptable, and coherent organizational structure. It offers a methodology to navigate the complexities of modern enterprises, integrating human ingenuity with digital potential and ensuring that organizations can thrive amidst constant change.

Ultimately, we can no longer construct large sociotechnical systems, of which enterprises are the best examples, without a grand design. This book presents such a blueprint, Unit Oriented Enterprise Architecture, inspired by the Fleet metaphor. This approach emphasizes the alignment of social, physical, and digital architectures, empowering the social core with technological extensions. The result is the creation of large sociotechnical systems composed of legitimate, purpose-driven, and accountable entities. These entities are proficient in executing essential human activities such as making, computing, and relating. However, the Unit Oriented Enterprise Architecture is not the only path. You are free to devise your own blueprint, provided it contributes to the preservation of our desired trajectory for human civilization—one that is effective, safe, and secure.

CHAPTER 9   CONCLUSION

Figure 9-1. *In Evolveburgh, strict by-laws require every book to have a Conclusion section. Local libraries, after numerous fines for noncompliance, now diligently write and glue in conclusions to every book, even cookbooks. Now, every recipe ends with "And they ate happily ever after."*

APPENDIX A

Evolveburgh—A City of Innovation and Tradition

The author expresses profound gratitude to the Mayor of Evolveburgh and the City Hall for their generous (albeit fictional) sponsorship and support. This appendix is dedicated to showcasing the fascinating and vibrant life of Evolveburgh, a city where the past and future intertwine seamlessly. Through these pages, may readers discover the charm, innovation, and unique character that make Evolveburgh a beacon of progress and tradition.

Acknowledgments and Gratitude

First and foremost, I would like to extend my deepest gratitude to the Mayor of Evolveburgh and the City Hall for their generous sponsorship and unwavering support. This appendix is a small token of my appreciation for the opportunity to illustrate the fascinating life of Evolveburgh.

APPENDIX A   EVOLVEBURGH—A CITY OF INNOVATION AND TRADITION

The Heart of Evolveburgh

Evolveburgh is not just a city; it's a living, breathing embodiment of the harmonious blend of tradition and cutting-edge innovation. As the second capital of Aitopia, Evolveburgh stands at the forefront of technological advancement while preserving its rich cultural heritage.

Evolveburgh's origins are steeped in a rich kaleidoscope of serendipitous events rather than meticulous planning. Founded at the confluence of several historical happenstances, the city emerged not through deliberate design but through a series of fortuitous circumstances that shaped its unique character. Over the centuries, Evolveburgh evolved organically, absorbing influences from various cultures, innovations, and epochs. This eclectic mix has resulted in a city that defies conventional definitions, existing as a living, breathing entity constantly reinventing itself.

Figure A-1. *Evolveburgh's scientists have uncovered that their city is an anomaly; it's not situated at the juncture of past and future but exists in a perpetual state of possibility. Evolveburgh is the Schrödinger's cat of cities—both everywhere and nowhere until observed*

APPENDIX A   EVOLVEBURGH—A CITY OF INNOVATION AND TRADITION

Thriving on the city's unconventional spirit, evident in its whimsical approach to urban development and community engagement, Evolveburgh's eco-friendly transit system has taken a novel turn with the introduction of pogo stick lanes. These lanes come complete with traffic lights that bounce up and down, stop signs you have to jump over, and a commuter loyalty program that rewards you with free springs for every 100 hops. The streets of Evolveburgh are also a testament to its artistic flair, featuring 3D crosswalks that create the illusion of floating above the ground, serving as both traffic calming measures and public art installations. Architecturally, the city has embraced a trend of constructing Pisa-style buildings with foundations skyward, captivating tourists and locals alike with their gravity-defying designs.

Entwined with the enigmatic presence of Schrödinger, Evolveburgh boasts a multitude of landmarks that celebrate its unique bond with the famed physicist. Visitors can marvel at the statues of Schrödinger—nothing too grand, just life-sized bronze figures, one of him, one of each of his relatives, and one of each of his pets, scattered throughout the city as if placed by the hand of uncertainty itself.

Figure A-2. *The town square features a statue of Schrödinger himself—less famous than his cat but not less entangled in Evolveburgh's history—honoring the city's quantum mechanics club, which meets every other now and then, in a superposition of times*

APPENDIX A   EVOLVEBURGH—A CITY OF INNOVATION AND TRADITION

The Quantum Café stands as a culinary tribute, where Schrödinger was rumored to savor meals that were both delicious and yet to be tasted, famous for its "Uncertainty Pie," which might be apple... or cherry—you'll only know once you take a bite. Art enthusiasts are drawn to the Wave-Particle Duality Gallery, a modest art space above the post office where each painting is a visual conundrum, simultaneously complete and awaiting its final stroke. The Schrödinger Auditorium, once the old town hall meeting room, is where Schrödinger himself might have discussed quantum mechanics with anyone who would listen. The Schrödinger Zoo is a place where all animals are regarded as both domesticated and wild until a visitor steps into the enclosure—though it's best not to inquire about the cat exhibit. Lastly, there's the Quantum Mechanics Club, an unpretentious little spot tucked away behind the library. It's where local enthusiasts gather to discuss the quirks of quantum theory over cups of strong coffee and homemade cookies. The club, with its mismatched furniture and walls lined with chalkboards, is a place of chaotic comfort. Members might engage in debates over equations or exchange jokes about particles positioned indeterminately—just don't anticipate any monumental discoveries, unless it's the occasional standout recipe for quantum quiche.

APPENDIX A   EVOLVEBURGH—A CITY OF INNOVATION AND TRADITION

Figure A-3. *The Relativity Square of Evolveburgh features the statue of Gottfried Wilhelm Leibniz, the last man who knew everything. Deep in thought, he's completely unaware of Albert Einstein approaching from two hundred years later, probably to compare notes on the theory of relativity—or perhaps just to ask for directions*

For history enthusiasts, the city boasts time capsule parks that transport visitors to different historical eras, featuring Evolveburgh's famous Time-Travel Trampoline, the Retro-Revolution Roundabout, and the Epoch Elevator. These attractions offer a unique blend of education and entertainment, allowing people of all ages to leap through the ages, spin back to bygone days, and ascend to significant milestones of human achievement. Whether you're bouncing into the future or circling the past, Evolveburgh ensures every visit is a journey through time itself.

APPENDIX A   EVOLVEBURGH—A CITY OF INNOVATION AND TRADITION

The city roads are so smart that visitors ask if they have PhDs in traffic management. With lanes that change direction as if they're indecisive chameleons, you'll never be stuck in a jam—unless it's the fruit kind, and even then, the city's drones will airlift you out.

The electricity grid in Evolveburgh is a silent powerhouse of efficiency. Streetlights here don't just illuminate; they're equipped with sensors that dim or brighten based on pedestrian traffic—making them not only smart but also considerate citizens of the city.

Public transportation is the city's pride, with a network of buses and trains that are so punctual, residents set their watches by them. The buses are hybrids, whispering down the lanes, and the trains glide on the rails like swans on a lake, all while maintaining an eco-friendly footprint.

On a negative side, Evolveburgh's roundabouts have become so efficient and enjoyable that they've inadvertently created a unique problem. Drivers find themselves circling for hours—mesmerized by the seamless flow—only venturing out for fuel. This phenomenon has led to an unexpected trend: residents are selling their homes, as the roundabouts have become their preferred dwelling place.

APPENDIX A EVOLVEBURGH—A CITY OF INNOVATION AND TRADITION

Figure A-4. *Evolveburgh's budget meeting got a little more interesting when the mayor tried to explain the $10-billion "Squaring the Endless Vicious Circles" line item. It turns out they're investing a small fortune to finally fix those roundabouts—by literally squaring them! Who knew geometry could be so expensive?*

To address this, the city is tackling the unexpected issue of roundabouts being too appealing by adopting a strategy of angularization—essentially, triangulating or squaring the roundabouts. Evolveburgh's plan involves reshaping the roundabout experience to encourage timely exits, thus preventing drivers from treating them as alternative homes. This angular approach aims to add a new dimension to the roundabout's design, making the exits more prominent and less likely to be bypassed for another loop. With a budget of $10 billion, the city hopes to redefine the roundabout, turning a potentially endless journey into a purposeful path that leads back home.

APPENDIX A EVOLVEBURGH—A CITY OF INNOVATION AND TRADITION

The Vibrant Community

Evolveburgh is not just about buildings and technology; it's about the people who make it vibrant and dynamic. The city's unique character is a testament to its ability to embrace paradoxes and contradictions. Evolveburgh is a mosaic of old and new, where ancient cobblestone streets coexist with cutting-edge technology hubs. Its residents are equally diverse, blending traditional values with avant-garde thinking. This dynamic interplay creates a vibrant atmosphere where innovation thrives amidst historical echoes.

Although the confluence of traditional values and avant-garde thinking typically harmonizes into a symphony of progress, there are moments when this fusion falters, and the melody of modernity clashes with the rhythm of the past. In these rare instances, the city's vibrant atmosphere is charged with a tension that is as palpable as the electric air before a storm.

Imagine a scenario where the city's annual heritage festival, a celebration steeped in tradition, is suddenly infused with a wave of futuristic art installations. The cobblestone streets, usually lined with historical reenactments, are overtaken by holographic displays projecting scenes from potential futures. The clash is immediate and jarring—the elders of Evolveburgh feel a dissonance, a rift in the very essence of their identity.

Yet, it is within this dissonance that Evolveburgh finds its true strength. For it is not in seamless blending but in the courageous confrontation of contrasts that the city's spirit is tested and tempered. The unexpected twist comes when, instead of rejecting the avant-garde intrusions, the residents find a new tradition in the making. They begin to weave these threads of the future into the fabric of their heritage, creating a mosaic that tells not just of where they have been, but of the infinite possibilities that lie ahead.

APPENDIX A   EVOLVEBURGH—A CITY OF INNOVATION AND TRADITION

Evolveburgh isn't just a place; it's an experience—an ever-changing landscape where the impossible becomes possible and the mundane transforms into the extraordinary. The citizens greet each other with "Deja vu!" as a testament to the city's unique charm, where every encounter is so remarkable, it feels like a rediscovery of wonder and delight.

Figure A-5. *In Evolveburgh, time is a question mark, the calendar is just a roulette wheel, the biggest annual event is the "Festival of Maybe," and standard greetings are "Deja vu!" It's the only place where you can have nostalgia for the future and anticipation for the past*

In Evolveburgh, communication is an art form, a performance that unfolds on the grand stage of daily life. The city's tech titans, visionaries in their own right, have long championed the charm of linguistic diversity, transforming what was once considered broken English into a celebrated mode of expression. With the recent introduction of the Accent Amendment, the very essence of verbal exchange has undergone a dramatic transformation. The decree is unequivocal: every utterance is a theatrical act, a display of one's unique vocal signature.

APPENDIX A   EVOLVEBURGH—A CITY OF INNOVATION AND TRADITION

Figure A-6. *The tech titans of Evolveburgh have long celebrated the quirks of broken English, elevating it to official status. However, the introduction of the Accent Amendment has seen a seismic shift in verbal communications. The rule is clear: to speak is to perform, and a neutral accent is now a relic of the past. The citizens, always quick on the uptake, has collectively tuned to the melodic lilt of Gibberish, ensuring compliance with the absurdity of existence and adding a touch of creative chaos to every interaction*

However, when the conversation becomes too theatrical, when the flamboyant flourishes of Gibberish eclipse clarity, the citizens of Evolveburgh have a failsafe. In these moments, they resort to the Silent Discourse—a communication form that transcends the spoken word. With a repertoire of expressive gestures, knowing glances, and the subtle art of telepathy via their intuitive tech wearables, they convey volumes without uttering a single syllable. This silent ballet of communication ensures that even in the midst of linguistic exuberance, the essence of the message is never lost. It's a reminder that in Evolveburgh, sometimes the most profound conversations are those that are felt, not heard.

Education in Evolveburgh

Evolveburghers take immense pride in their educational institutions, which are renowned for their commitment to maintaining the purity of scientific inquiry, free from the taint of alchemy and other pseudosciences. The city's universities are bastions of rigorous scholarship, dedicated to fostering genuine intellectual growth and innovation.

The Universities of Evolveburgh

Evolveburgh is home to several prestigious universities, each contributing uniquely to the city's rich academic landscape. The universities uphold a high standard of education, emphasizing critical thinking and empirical research. Their commitment to keeping sciences pure has earned them a global reputation for excellence.

> **Retro-Futurism College (RFC):** At RFC, the past and the future collide in the study of Retro-Futurism. Scholars here are dedicated to reviving the technologies of yesteryear for tomorrow's world. Their crowning achievement is the Steam-Powered Smartphone, a device that combines Victorian aesthetics with modern functionality, complete with a brass gear-operated selfie stick.

Mythological Sciences School (MSS): This institution is the epicenter of Mythological Zoology. The faculty and students at MSS are on a quest to prove the existence of mythical creatures. Their most ambitious project is breeding a unicorn, which, according to their research, is just a horse that has been misinformed about the horn situation.

Paranormal Psychology Institute (PPI): PPI explores the intersection of the mind and the mysterious in Paranormal Psychology. Here, they investigate whether ghosts suffer from existential crises and if poltergeists are simply misunderstood. Their latest seminar series, "Hauntings and the Human Psyche," seeks to offer therapy sessions for troubled spirits.

Figure A-7. It remained a mystery why every graduate from Evolveburgh's least prestigious "Oops" University won Nobel Prizes, until their janitor published his pamphlet, "The Quantum Serendipity Effect." The pamphlet became a bestseller overnight and earned him the Nobel Prize for Literature. At the award ceremony, he spilled the beans: it was all thanks to the real brainiac, the university mascot, and their "enlightening" silent debates

APPENDIX A   EVOLVEBURGH—A CITY OF INNOVATION AND TRADITION

Maintaining the Sanctity of Scientific Inquiry

Elite universities employ several strategies that, while unorthodox, serve as a testament to Evolveburgh's commitment to upholding the highest standards of scientific integrity, ensuring that its graduates are well equipped to contribute to the global body of knowledge:

- **The Evidence-Based Bootcamp**: Before embarking on their research journey, all incoming graduate students must complete the Evidence-Based Bootcamp. This intensive program drills into them the importance of empirical data and statistical rigor. They learn to question every assumption, scrutinize every data point, and leave no hypothesis untested. It's a rite of passage that instills in them the discipline required to conduct research that's free from the taint of pseudoscience.

- **The Methodological Maze**: Another unique strategy is the Methodological Maze—a literal labyrinth that doctoral students must navigate. Each twist and turn of the maze corresponds to a step in the scientific method, from hypothesis formation to data analysis. This physical embodiment of the research process serves as a powerful metaphor, emphasizing that there can be no shortcuts in the pursuit of truth.

- **The Peer Review Gauntlet**: Finally, to combat the spread of pseudoscience, the universities have established the Peer Review Gauntlet. Here, scholars must defend their work against a panel of their most critical peers, armed with the sharpest of scientific scrutiny. This intense process ensures that only ideas that can withstand the most rigorous of challenges are allowed to enter the academic canon.

APPENDIX A   EVOLVEBURGH—A CITY OF INNOVATION AND TRADITION

- **The Integrity Inquisition**: A panel of seasoned scholars, known as the Integrity Inquisition, oversees the adherence to academic honesty. They conduct random audits of research work and provide guidance on ethical scholarship. Any student or faculty found in violation of the universities' strict plagiarism policies faces the Inquisition's swift and fair judgment, ensuring the purity of Evolveburgh's academic endeavors remains untainted.

Figure A-8. *To prevent the emergence of speculative theories, Evolveburgh's elite universities now mandate that graduates defending their dissertations must physically stand on the shoulders of giants. If you want your PhD, better practice your balance—these human pillars offer a challenging platform!*

360

Commitment to Pure Sciences

Evolveburgh's dedication to pure sciences is not just a tradition but a way of life. The city's academic community works tirelessly to advance knowledge while safeguarding it from pseudoscience. This unwavering commitment ensures that Evolveburgh remains at the forefront of scientific innovation, producing scholars who are both grounded in rigorous methodology and inspired by the limitless possibilities of discovery.

Through their unique practices and unwavering dedication, Evolveburgh's universities embody the city's spirit of intellectual curiosity and excellence, making it a beacon of true scientific pursuit.

Business in Evolveburgh

Evolveburgh is not only a hub of academic excellence but also a bustling center of innovative business ventures. The city's unique blend of tradition and futuristic thinking creates a fertile ground for entrepreneurial endeavors that push the boundaries of conventional business practices.

Cutting-Edge Ventures

Evolveburgh's business landscape is characterized by its daring and sometimes unconventional ventures. Entrepreneurs here are known for their willingness to explore the unknown, often embarking on projects that seem outlandish yet hold significant potential.

The Televeterinary Titan

Spearheaded by the mayor's second cousin, who once mistook a raccoon for a cat, this venture aims to revolutionize pet care. The project, "Paws and Play," offers a virtual reality experience for diagnosing pets, claiming that animals respond better to avatars than actual vets. The service includes a VR headset for your furry friend, ensuring they remain blissfully unaware while being virtually poked and prodded.

The Aquatic Agriculture Adventure

The brainchild of the mayor's yoga instructor, who believes in the healing power of water, this venture, "HydroHarvest," explores the untapped potential of underwater farming. The project aims to grow organic vegetables on the ocean floor, harnessing the nutrients from marine ecosystems. Critics are skeptical about the practicality of scuba diving for your salad, but proponents argue it's a deep dive into sustainable agriculture.

Interstellar Ventures

One of the most ambitious projects currently underway involves exploring interstellar resources. This project, led by the mayor's son-in-law, aims to utilize not only advanced technology but also his advanced relationships with the authorities for groundbreaking returns. While experts hail it as a significant opportunity with unprecedented potential, critics question the ethics, suggesting the investment might be disappearing into a financial void or somebody's pockets.

APPENDIX A   EVOLVEBURGH—A CITY OF INNOVATION AND TRADITION

Figure A-9. Breaking News: Evolveburgh's first interstellar venture hits a "dark" patch as mayor's son-in-law seeks a modest $215 billion to mine Bitcoin from the nearest black hole. Experts say it's a "stellar" opportunity, but critics question if money is just being sucked into a financial void. Stay tuned for updates on this "gravity" of the situation!

Financial Time Travel

In another groundbreaking initiative, Evolveburgh has successfully developed a financial digital time machine aimed at reversing economic crises. This innovative device has the potential to undo financial missteps and restore economic stability, showcasing the city's commitment to

APPENDIX A   EVOLVEBURGH—A CITY OF INNOVATION AND TRADITION

forward-thinking solutions. However, the unexpected side effects of such powerful technology highlight the unpredictable nature of pioneering business ventures in Evolveburgh.

Figure A-10. *Evolveburgh's efforts to develop a financial digital time machine were crowned with relative success. The machine reversed the city's financial crisis, erasing not only residents' losses but also, as an unwelcome side effect, the mayor's secret Bitcoin stash. Turns out, the vintage Ponzi scheme and cryptoboomerang were more than just decorative pieces on the machine's motherboard!*

APPENDIX A   EVOLVEBURGH—A CITY OF INNOVATION AND TRADITION

Entrepreneurial Ecosystem

In Evolveburgh, where the entrepreneurial spirit soars higher than the city's famed skyscrapers, a robust ecosystem stands as the bedrock of innovation and prosperity. Here, the term "ecosystem" is not just a fancy buzzword but a living, breathing entity that nurtures the seeds of business with a touch that's as organic as it is strategic. When envisioning Evolveburgh's entrepreneurial ecosystem, readers should imagine not just caves, but an entire subterranean wonderland. Think of it as a vast network of interconnected caverns, each echoing with the whispers of commerce and the clink of coin.

- **Bioluminescent forests**: Imagine walking into a cave and finding a forest glowing with the light of bioluminescent plants. These are the brainstorming groves, where ideas grow like mushrooms in the dark, fed by the rich soil of collaboration and creativity.

- **Crystal clear streams**: Flowing through these caves are streams of crystal-clear water, symbolizing the transparent communication channels that keep the ecosystem thriving. These waters carry knowledge and insight, quenching the thirst for innovation.

- **Stalactite boardrooms**: Look up, and you'll see stalactites forming natural boardrooms. Here, the stalwart leaders of industry meet, their decisions shaping the contours of the business landscape like the slow, steady drip of mineral-rich water.

- **Geothermal springs**: Pockets of geothermal energy represent the fiery passion of entrepreneurs, providing warmth and power to drive the engines of growth. These springs are the hotbeds of activity, bubbling with potential and new ventures.

- **Fossilized success stories**: Embedded in the walls are the fossilized remains of successful startups from eons past, serving as inspiration and a reminder that with the right conditions, life in business can thrive and leave a lasting impact.

In this subterranean realm, the spirit of entrepreneurship is not just incubated; it's a force of nature, carving out its own path through the bedrock of the market. It's a place where the term "ecosystem" truly comes to life, complete with its own natural wonders and a sense of mystery that beckons the bold to explore its depths. Evolveburgh is not just a city; it's an adventure for the business-minded, a journey into the heart of innovation itself.

Ethical Business Practices

Evolveburgh places a strong emphasis on ethical business practices. The city's regulatory framework is designed to prevent corporate malfeasance and ensure that businesses operate with integrity. Several specialized committees have been established, each focusing on different aspects of business ethics:

- **Ethical Practices Committee**: Headed by the mayor's step-aunt's best friend's dog walker, this committee oversees the implementation of ethical guidelines and provides training to ensure all businesses adhere to the highest standards of integrity. Their mandate includes regular audits and workshops to promote ethical decision-making across the corporate sector.

- **Anti-nepotism Committee**: Chaired by the mayor's favorite barista, this committee is dedicated to preventing nepotism within Evolveburgh's business community. They conduct thorough investigations into hiring practices and enforce policies that promote merit-based advancement, ensuring fair opportunities for all.

APPENDIX A EVOLVEBURGH—A CITY OF INNOVATION AND TRADITION

- **Transparency and Accountability Committee:** Led by the mayor's neighbor's psychic's childhood friend, this committee focuses on enhancing transparency within businesses. They require companies to disclose financial records and decision-making processes, thereby fostering an environment of openness and trust.

- **Corporate Social Responsibility Committee:** Managed by the mayor's cousin's hairdresser, this committee encourages businesses to engage in socially responsible practices. They promote initiatives that benefit the community and environment, ensuring that corporate actions align with societal values.

- **Anti-corruption Task Force:** Under the leadership of the mayor's childhood pen pal, this task force aggressively tackles corruption within the business sector. They implement strict anti-corruption measures, investigate allegations, and ensure that any breaches are met with swift and fair consequences.

- **The Integrity Inquisition for Business:** This formidable panel, consisting of seasoned scholars and industry experts, conducts audits and provides guidance on best practices. They scrutinize business operations to ensure compliance with ethical standards and maintain the high ethical reputation of Evolveburgh's corporate landscape.

The recent investigation into Evolveburgh's business community has shed light on the operations of Grave Concerns, a funeral company owned by the mayor's aunt, Mary. Contrary to the claims made by some journalists, the investigation revealed that Grave Concerns has been conducting its affairs with impeccable transparency and adherence to ethical standards.

APPENDIX A   EVOLVEBURGH—A CITY OF INNOVATION AND TRADITION

In a definitive response to the allegations, it was established that Grave Concerns is not in debt to the city for $6.5 million. Instead, it is the city that finds itself indebted to the company, owing a substantial sum of $14 million. This revelation not only clears Grave Concerns of any financial misconduct but also highlights the company's robust and principled business practices.

Figure A-11. In a display of both vindication and wit, Grave Concerns has aptly incorporated "Crystal Clear Mannerism" into their logo, symbolizing their commitment to clarity and honesty in their dealings. This development serves as a testament to the integrity of Evolveburgh's businesses and stands as a correction to the erroneous reports previously circulated. Grave Concerns' reputation remains untarnished, and the city's business environment continues to be characterized by its transparency and ethical conduct

APPENDIX A   EVOLVEBURGH—A CITY OF INNOVATION AND TRADITION

In Evolveburgh, ethical business isn't just a practice; it's a family affair that ensures the business environment remains ethical, transparent, and fair. This multifaceted approach reflects the city's unwavering commitment to upholding the highest standards of business integrity.

Politics in Evolveburgh

In Evolveburgh, the political landscape is as stable as a table with three legs, one of which has decided to pursue a career in modern art, standing awkwardly as a metaphor for the delicate balance of democracy. Every citizen is a self-declared prime minister in waiting, contributing to a vibrant yet perplexing tapestry of governance.

APPENDIX A   EVOLVEBURGH—A CITY OF INNOVATION AND TRADITION

Figure A-12. *Contrary to the grand myth of Earth standing on three titanic whales, Evolveburgh's top minds now believe Earth stands on three Schrödinger's cats, each in a state of perpetual uncertainty—just like our political climate. Stability? It's a quantum question*

The mayor, initially elected on a platform championing transparency and efficiency, has encountered a shift in responsibilities. The focus has transitioned from the implementation of policies to providing clarifications on the city council's unconventional decisions. This council, characterized by its diverse composition of individuals with philosophical inclinations and others of varying intellectual repute, convenes biweekly to deliberate on matters of civic significance. Their discussions often revolve around seemingly minor, yet symbolically significant, urban details such as determining the ideal dimensions for public benches. This meticulous attention to detail reflects a broader commitment to enhancing the quality of communal spaces.

Figure A-13. As the council members debate the optimal length, width, and existential purpose of a bench, one can't help but admire their dedication. After all, it's not just a bench; it's a symbol of rest, reflection, and the occasional bird visitation. And so, the council continues, ever committed to the noble cause of perfecting the town's sitting experience, because in Evolveburgh a bench is where ideas take flight, not just the birds

Furthermore, the council engages in reflective discourse on proverbial wisdom, exemplified by their analysis of the mayor's expression, "A bit for your questions, but a coin for your paradoxes." This phrase suggests that while questions may be common, it's the paradoxes that hold true value, much like the elusive Satoshi Nakamoto's Bitcoin—a currency of insight in the economy of governance. Such deliberations serve to evaluate the efficacy of collective decision-making processes within municipal governance. In this context, the mayor's role has evolved into

APPENDIX A   EVOLVEBURGH—A CITY OF INNOVATION AND TRADITION

that of an intermediary, tasked with interpreting the council's avant-garde resolutions into practical, executable strategies. This development underscores the dynamic nature of public service roles and the adaptability required to fulfill them effectively.

In Evolveburgh, the political landscape is marked by a vibrant opposition that plays a crucial role in the city's governance. Opposition parties, though diverse in their ideologies, share a common goal: to scrutinize the actions of the ruling party, ensure accountability, and foster transparency, all while maintaining a flair for dramatic speeches and exaggerated gestures. Their presence is not merely symbolic; it is instrumental in promoting a culture of good governance and serving as a credible alternative to the current administration—when they're not busy arguing over the best brand of coffee for council meetings.

As the campaign season in Evolveburgh approaches, the political atmosphere intensifies, and the air is thick with promises and the scent of freshly printed flyers. Candidates from various parties embark on a rigorous journey to connect with citizens, articulate their platforms, shake hands, kiss babies (metaphorically speaking, of course), and win hearts with their grand plans for the city's future.

The campaign trail is a mosaic of town hall meetings, where the coffee is strong and the opinions stronger, policy debates that sometimes resemble a polite battle of the bands, and community outreach programs where every handshake is a silent plea for votes. It's a dance of democracy, and everyone's invited to tango.

Meanwhile, the opposition is playing a game of civic chess, moving pieces that represent their multifaceted strategy. They're the challengers, the underdogs, the ones pointing fingers at the incumbents with a "We can do better" twinkle in their eyes. Their campaign is a mixtape of immediate fixes and visionary blueprints, all wrapped up in a slogan that's both catchy and slightly vague.

APPENDIX A   EVOLVEBURGH—A CITY OF INNOVATION AND TRADITION

Figure A-14. *The opposition's activities extend beyond the council chambers, actively engaging with the community to stimulate public debate and serve as a training ground for future leaders. They often host workshops for five-year-old aspiring mayors on advanced eye-rolling techniques, such as the "Classic Disapproval," the "Sarcastic Surprise," and the "Why-Did-I-Even-Bother" eye-roll. Because nothing says "future leader" quite like a well-timed eye-roll*

They tackle urban planning with the promise of more roundabouts than you can reasonably navigate, economic growth with plans that are sure to multiply jobs like rabbits, and social welfare with a warmth that could melt the coldest of bureaucratic hearts. It's election season in Evolveburgh, and the competition is as intense as a prisoner's dilemma at a twin convention—everyone looks the same, promises are identical, and voters are left wondering who's the evil twin!

APPENDIX A   EVOLVEBURGH—A CITY OF INNOVATION AND TRADITION

Enterprise Architects in Evolveburgh

In Evolveburgh, enterprise architects are more than a demographic—they're practically a cultural phenomenon, representing roughly 20% of the population. Each one is unique, holding titles as varied as Chief, Distinguished, Principal, or Senior, scattered throughout city departments and private firms alike. Despite their differences, they share a curious hallmark: an unyielding allegiance to a personal version of reality.

The enterprise architects of Evolveburgh are perpetually lost in the realm of their thoughts, making it nearly impossible to get their attention. So absorbed are they in their own realities that they often fail to notice anything outside them. The Chief Architect, however, stands apart; he alone acknowledges the existence of multiple realities yet chooses to stay rooted in the one where documentation is always up to date. Watching over the chaotic multiverse from his perfect world, he ensures that every document remains pristine and that chaos is but a distant memory.

These architects have their quirks. Elusive and mysterious, they make catching a cloud seem easier than capturing their attention. Despite their busy demeanor, they've mastered the art of appearing occupied. To them, every document is a masterpiece in the making, dedicated to the grand blueprint that they believe shapes the "enterprise galaxy." Small talk is entertained only if it's mission-critical, and peripheral trivia has no place in their meticulously structured world.

Interacting with Evolveburgh's enterprise architects requires a touch of finesse. For instance, if you're a wolf in sheep's clothing, you have little to worry about. Their gaze is selective, drawn more to your fluffy exterior than to any lurking fangs beneath. In Evolveburgh, wolves disguised as sheep find sanctuary; just make sure your "baa" is pitch-perfect.

APPENDIX A   EVOLVEBURGH—A CITY OF INNOVATION AND TRADITION

Figure A-15. *Evolveburgh's enterprise architects have a knack for spotting wool, not wolves. So if you're a wolf in sheep's clothing, just perfect your baa and you'll blend right in*

And charisma? It has a peculiar effect on these architects. Faced with such presence, they transform into high-tech statues, pondering the mysteries of life—or perhaps just counting down to their next coffee break. Under the spell of charm, they take on their best impersonation of *The Thinker*, as if carved by Michelangelo himself.

Those who make the mistake of suggesting that Evolveburgh's enterprise architects might be less than likable quickly learn the price. Such hints act as social cloaking devices, ensuring that you vanish from their world entirely. They might not show it, but they take things literally. If you ask, "If you were in my shoes, how would you go about it?" don't be surprised if they step into your sneakers, exploring new career paths in your footwear. It's not uncommon to find them leading an impromptu theater troupe in borrowed shoes, turning the office into an unexpected stage.

Figure A-16. Pose the question, "If you were in my shoes," to enterprise architects, and before you know it, they're leading an improv theater group in your footwear, turning the office into a stage

Despite their eccentricities, Evolveburgh's enterprise architects are masters of the productive middle ground. They never expect to get what they want, and they know you won't, either. For them, playing the ultimate "compromise" game is a fascinating pastime. Here, nobody wins, but everyone enjoys the process of trying. They have perfected the art of low expectations, making you believe that your half-baked idea was their ingenious plan all along.

Their approach to problem-solving relies on the belief that the greatest scientific breakthroughs arise from daydreams or random tangents. Aware of this, their customers send them comfy pillows and random trivia books, then hide behind potted plants and office furniture, eagerly awaiting the next "eureka" moment.

Figure A-17. In Evolveburgh, no breakthrough is complete without a plush pillow, a trivia book, and a hidden audience holding its breath behind the office decor

APPENDIX A   EVOLVEBURGH—A CITY OF INNOVATION AND TRADITION

In Evolveburgh, new platforms are born from creative flashes of genius, with recent innovations moving from data lakes to data lakehouses, the Internet of Things, and now AI hallucinations. Next on the horizon? A new concept: "Data Mirage," the platform that almost exists.

To outsiders, Evolveburgh's enterprise architects might seem eccentric. But in this city, their quirks are woven into the very fabric of life, proving that in every plan, no matter how complex, there's always room for a little mystery.

Looking Forward

As we move into the future, Evolveburgh continues to set an example of how tradition and innovation can coexist harmoniously. The city seamlessly blends the old with the new, creating a vibrant tapestry that is both historically rich and technologically advanced. However, Evolveburgh's relationship with the future is anything but straightforward. At times, the city displays a sense of unnecessary complacency and even arrogance, confidently asserting its place at the forefront of progress.

Figure A-18. *When the future came knocking with its high-tech gadgets and grand promises, Evolveburgh replied, "Sorry, we're already light years ahead!"*

APPENDIX A EVOLVEBURGH—A CITY OF INNOVATION AND TRADITION

This complacency, however, is not rooted in the ill nature of its citizens. Instead, it can be traced back to several peculiar incidents involving visitors from the future who decided they preferred the present Evolveburgh and chose not to return to their own time. These instances have led to a significant issue of illegal immigration, with many of these future visitors—both human and alien—opting to stay in Evolveburgh, thereby complicating the city's demographic landscape.

Interestingly, while Evolveburgh faces challenges with illegal immigrants from the future, immigration from the past is virtually nonexistent. This is likely due to the absence of time machines that can facilitate such journeys. Only on rare occasions have immigrants from the future made brief stops in the past before settling in present-day Evolveburgh, creating a complex temporal mosaic.

Figure A-19. *Things in Evolveburgh were going so well that when somebody from the future came to visit, they asked for a return ticket—apparently, the future's not what it used to be!*

APPENDIX A   EVOLVEBURGH—A CITY OF INNOVATION AND TRADITION

Despite these challenges, the citizens of Evolveburgh maintain a remarkable optimism. There is a widespread awareness and preparation for potential doomsday scenarios, reflecting the city's unique blend of forward-thinking and pragmatic readiness. In Evolveburgh, the spirit of innovation is matched by a readiness for any eventuality, even the end of the world. This unique combination of progress and preparedness ensures that the city remains a fascinating and resilient place, ever ready to face whatever the future—or the past—might bring.

Figure A-20. Experience Evolveburgh: where the only thing evolving faster than the city is the countdown to Doomsday. Dress your best, dust off your history books, polish your monocle, and join us in Evolveburgh—where the apocalypse is just the grandest of finales! Remember, in Evolveburgh, every day could be your last—so party like there's no tomorrow, because there just might not be! Stay for the chaos, leave with a "Survived the End of Days" commemorative coin!

APPENDIX A   EVOLVEBURGH—A CITY OF INNOVATION AND TRADITION

In Evolveburgh, every day is an opportunity to celebrate the present while preparing for the future. Whether it's through their cutting-edge technology or their rich cultural traditions, the people of Evolveburgh are always looking forward, embracing the unknown with a blend of curiosity and caution. This spirit of adaptability and resilience is what truly sets Evolveburgh apart, making it a city like no other.

APPENDIX B

Bibliography

Bill, Max. Form: A Balance Sheet of Mid-Twentieth-Century Trends in Design. New York: Wittenborn, 1952.

Bughin, Jacques, Michael Chui, and James Manyika. "Ten IT-enabled Business Trends for the Decade Ahead." McKinsey Quarterly, May 2013.

de Greene, Kenyon B., ed. A Systems-Based Approach to Policymaking. New York: Springer Verlag, 2014.

Dunér, David, and Alan Crozier. The Natural Philosophy of Emanuel Swedenborg: A Study in the Conceptual Metaphors of the Mechanistic World-View. Dordrecht: Springer, 2013.

Feibleman, James K. "Theory of Integrative Levels." The British Journal for the Philosophy of Science 5 no. 17 (1954): 59–66.

Forrester, Jay W. "System Dynamics and the Lessons of 35 Years." In A Systems-Based Approach to Policymaking (pp. 199–240) edited by Kenyon B. de Greene. New York: Springer Verlag, 2014.

Gall, John. Systemantics: The Underground Text of Systems Lore: How Systems Really Work and Especially How They Fail. Ann Arbor, MI: General Systemantics Press, 1990.

Garvin, D. "Building a Learning Organization." Harvard Business Review, July–August 1993, pp. 78–91.

Giere, R. N. "How Models Are Used to Represent Reality." In PSA 2002: Proceedings of the 2002 Biennial Meeting of the Philosophy of Science Association. Part 2 (pp. 742–752), edited by Sandra D. Mitchell. Chicago, IL: University of Chicago Press, 2004.

Habraken, N. J., and Jonathan Teicher. The Structure of the Ordinary: Form and Control in the Built Environment. Cambridge, MA: MIT Press, 2000.

Heylighen, Francis. "Stigmergy as a Universal Coordination Mechanism: Components, Varieties and Applications." In Human Stigmergy: Theoretical Developments and New Applications (Studies in Applied Philosophy, Epistemology and Rational Ethics) (pp. 1–42), edited by Ted Lewis and Leslie Marsh. New York: Springer, 2015.

Juechter, W. Mathew. "Responding to Three Phases of Growth." In A Blueprint for Managing Change (pp. 15–20), edited by Joseph L. McCarthy. New York: Conference Board, 1996.

Klir, George J. "Complexity: Some General Observations." In Systems Research, vol. 2, no. 2, pp. 131–140. New York: Pergamon Press, 1985.

Kotter, John P. "Hierarchy and Network: Two Structures, One Organization." Harvard Business Review. Blog Network. https://hbr.org/2011/05/two-structures-one-organizatio.

Kransdorff, Arnold. Corporate DNA: Using Organizational Memory to Improve Poor Decision-Making. Aldershot, UK: Gower, 2006.

Lane, David Avra. "Emergence by Design." Kennisland, May 2011. https://www.kl.nl/wp-content/uploads/2014/04/md-b-ed..pdf.

Langer, Susanne K. An Introduction to Symbolic Logic. New York: Dover Publications, 2011 [1937].

Mak, Sander. "Modules vs. Microservices: Apply Modular System Design Principles While Avoiding the Operational Complexity of Microservices." O'Reilly, 2017. https://www.oreilly.com/radar/modules-vs-microservices/.

Malone, Thomas W., and Kevin Crowston. The Interdisciplinary Study of Coordination. Cambridge, MA: Center for Coordination Science, Alfred P. Sloan School of Management, Massachusetts Institute of Technology, 1993.

Matthews, Charles H., and Ralph Brueggemann. Innovation and Entrepreneurship: A Competency Framework. London: Routledge, 2015.

McCarthy, Joseph L. A Blueprint for Managing Change. New York: Conference Board, 1996.

Mey, Kerstin, and Simon Yuill. Cross-Wired: Communication-Interface-Locality. Manchester: Manchester University Press, 2004.

Milchman, Andre. AI Safety and Security: Architectural Context, Perspectives, and Insights. Toronto: Independently published, 2017.

Milchman, Andre. Enterprise Architecture Reimagined: A Concise Guide to Constructing an Artificially Intelligent Enterprise. Toronto: CreateSpace, 2017.

Milchman, Andre. The Triadic Evolution: Reimagine Tomorrow's Sociotechnical World. Toronto: New Key, 2024.

Milchman, Andre, and Fang Noah. Prescriptive Analytics: A Short Introduction to Counterintuitive Intelligence. Toronto: CreateSpace, 2018.

Milchman, Andre, and Fang Noah. The Transformable Enterprise: Enterprise Architecture in a New Key. Toronto: New Key, 2021.

Rickards, Tudor. Creativity and Problem Solving at Work. Aldershot: Gower, 1997.

Rudström, Åsa, Kristina Höök, and Martin Svensson. "Social Positioning: Designing the Seams between Social, Physical and Digital Space." In Proceedings of the 1st International Conference on Online Communities and Social Computing. Las Vegas: HCII, 2005.

Sadik-Khan, Janette, and Seth Solomonow. Streetfight: Handbook for an Urban Revolution. New York: Penguin Books, 2017.

Sessions, Roger. "Interview with John Zachman." Perspectives of the International Association of Software Architects, 2007.

Sewell, Marc T., and Laura M. Sewell. The Software Architect's Profession: An Introduction. Upper Saddle River, NJ: Prentice Hall PTR, 2002.

Thackara, J. "Foreword." In The Handbook of Design for Sustainability, edited by Stuart Walker, Jacques Giard, Helen L. Walker, xxv. New York: Bloomsbury, 2013.

APPENDIX B   BIBLIOGRAPHY

van Fraassen, Bas C. Scientific Representation Paradoxes of Perspective. Oxford: Clarendon Press, 2013.

Wallech, Steven. China and the West to 1600: Empire, Philosophy and the Paradox of Culture. Malden: Wiley Blackwell, 2016.

Wiggins, A. "The Twelve-Factor App." 2017. https://12factor.net/.

Zachman, John. 1982. Business Systems Planning and Business Information Control Study: A Comparison. IBM Systems Journal vol. 21, no. 1, pp. 31–53 (1982).

Zachman, John. "Defining Enterprise Architecture: The Systems Are the Enterprise." 2015a. https://www.zachman.com/resources/zblog/item/defining-enterprise-architecture-the-systems-are-the-enterprise.

Zachman, John. "Enterprise Architecture is an Enterprise Issue, NOT an IT Model-Building Exercise." 2015b. https://www.zachman.com/resources/zblog/item/enterprise-architecture-is-an-enterprise-issue-not-an-it-model-building-exercise.

Zachman, John. "Intro to EA: The Paradigm Problem." 2015c. https://www.zachman.com/resources/zblog/item/intro-to-ea-the-paradigm-problem.

Zachman, John. "The Issue is THE ENTERPRISE." 2015d. https://www.zachman.com/resources/zblog/item/the-issue-is-the-enterprise.

Index

A

Ability (physical capability of humans), 56–58
Ability to recycle digital resources, 60
Accountability, 75, 76, 251, 264, 342
Accountability within Evolutionary Units, 109
Adaptability, 318
Adaptation, 95
Adaptive systems, 50
Agent, 217
Alignment of Boundaries, 77
Angularization, 353
Anti-corruption, 367
Anti-nepotism committee, 366
Application Programming Interfaces (APIs), 306
Applications and limitations, 52
Architectural approach, 251
    characteristics, 18
    comprehensibility, 170
    positioning components, 170, 172
    relating components, 170, 173, 174
    selecting components, 170, 171
    specific needs, 17

Architectural elements uniformity and consistency, 178–180
Architectures
    style, 341
    types, 25–27
Armada
    formation of ships, 205
    organizational structure and functions, 207
    standardization and flexibility, 208
Armada metaphor, global enterprise, 200, 201
Asynchronous communication, 313, 314, 324, 338, 339
Automated service interfaces, 304, 305
Autonomy, 77, 78, 265
AWS Direct Connect, 124
Azure ExpressRoute, 124
Azure implementation, 284–286

B

Basic Star topology, 229
Bifocal approach, 252
Bifocal design approach, 267

INDEX

Biological evolution, 90, 91
Blueprint for accountable
    enterprise, 249
Building blocks, 254, 255
Business in Evolveburgh
    cutting-edge ventures, 361–369
    innovative business, 361
Business Process Management
    (BPM), 314, 315
    services, 34
    software, 33
Business-Technology Alignment, 210
Bus structure, 181

C

Cargo, 206
Central control and local
    autonomy, 230
Centralized organization and
    coordination
    mechanism, 30
Changeability, 170
Civilization, 115
Clear ownership and control, 210
Cloud connectivity, 124
Cloud structuring, 290
Code organization, 290
Coevolution, 96, 97
Coherence in Purpose
    Implementation, 219, 220
Cohesive modeling (code
    based), 139
Collaboration, 232

Collaborative and cooperative
    relations, 79
Collaborative relations, 190
Command-and-Control, 30–33
Communication, 357
Communication in sociotechnical
    systems, 122
Community relations, 81
Competing stakeholders, 11
Competitive interactions, 320–322
Competitive relations, 80
Complex systems, 140, 144
Compliance interactions, 318, 319
Compliance relations, 80, 190, 233
Complicated dependencies among
    business units, 28
Component and modular
    approaches, 73–75
Composability, 77
Composite modeling (textual), 139
Compositional pattern(s), 249
    distributed composition, 209
Comprehensibility, 170, 177
Comprehensive
    blueprints, 134–136
Configurability, 49
Connective modeling
    (metaphoric), 139
Consumer expectations, 307
Consumer markets, 222
Containerization approach, 206
Contemporary enterprises
    physical, digital, and social
    dimensions, 24, 25

Contemporary technology, 18
Content management system (CMS), 51
Conversational interfaces, 306
Cooperative interactions, 321, 322, 336, 337
    elements, additional, 323, 324
    mechanisms, 324–326
Cooperative relations, 190, 232
Coordinating collaborative interactions
    asynchronous communication methods, 316
    human coordination, 314, 315
    indirect coordination, 313, 314
    pre-designed structures, 317–319
    stigmergy, 315, 316
Coordinating organizational interactions
    directive interfaces, 309
    leveraging technology, 312
    reporting interfaces, 311
    structured business rules interfaces, 309, 310
    types of interfaces, 308
    unstructured documents and guidelines, 310, 311
Coordination, 88, 115, 232
    communication and connectivity, 121, 122
    within organizations, 116
Coordination, forefront of enterprise operations and strategy, 237
Core integrative component, 184
Corporate Social Responsibility Committee, 367
Craft, 215
Crew, 215
Cross-cloud data management, 125
Culture constructs, 261
Cutting-edge technology, 381
Cutting-edge ventures
    aquatic agriculture adventure, 362
    entrepreneurial ecosystem, 365, 366
    ethical business practices, 366–369
    financial time travel, 363, 364
    interstellar ventures, 362
    televeterinary Titan, 361

D

Data protection, 4
Data security, 308
Data Sharing, 232
Decentralization, 226
Decision-making and integration, 177
Decisive leadership, 10
Demonstration (how), 137
Demystification of Enterprise Architecture, 235, 236
Denotation (what), 137, 138
Departmental segmentation, 278

Designing units
  purpose-driven constructs, 253, 257–262
  structural uniformity, 252, 254–257
DevOps, 341
Digital advancements, 1
Digital capabilities (ability to compute), 58–60
Digital constructs, 258, 259
Digital crafts, 266, 267
Digital enterprise
  army, 160–162
  city, 150–152
  coordinated fleet, 162–165
  factory, 157–159
  hospital model, 153–157
  invisibility, 7
  interpretations and design pathways, 156
  University, 157
Digital environments
  life cycle, 21–23
Digital systems post-acquisition, 10
Digital Within Social Legitimacy, 218
Direct coordination, 118, 119
  advantages, 328
  disadvantages, 328
  key attributes, 327, 328
Directional integration, 78
Directionality (where), 137
Direction-Setting, 233
Directive interfaces, 309

Disaster struck at Evolveburgh's grand premiere, 11
Distributed architecture and loose coupling, 225
Distributed composition, 194, 195, 199
Diversity within evolutionary stream, 96
Double nesting, 230
Dynamics, 50

E

Education in Evolveburgh
  commitment, 361
  scientific inquiry, 359, 360
  Universities, 357, 358
Email, 34, 36
Emergence properties, 49
Employment relations, 191
Empowerment, 76
Empowerment through possession and control, 110
Encapsulation, 76
Engineering, 274–276
  applications as apartments, 288
  applications as rooms, 287, 288
  uniform structure, 287
  vendor-provided solutions, 289
Engineering and design contexts, 135
Enhanced systems thinking, 97
Enterprise architects, 7, 15, 374–378

INDEX

Enterprise architecture, 342
  adaptability and
    responsiveness, 44
  architectural-centric
    approach, 71
  characteristics of systems
    non-structural
      characteristics, 49, 50
    structural
      characteristics, 48, 49
  clarity and
    comprehensibility, 71
  communicable approach, 44
  critical information and slow
    response times, 44
  demystification, 235, 236
  design methodology, 43
  holistic and unit-oriented
    approach, 83
  holistic view, 72
  human-centric needs, 82
  information exchange, 44
  integration and functionality, 71
  knowledge silos and
    communication
    barriers, 43
  organic structures, 53
  pressing needs, 43
  sociotechnical systems, 43, 44
  strategic and operational
    objectives, 71
  strategic integration of entire
    systems, 83
  streamlined communication, 44
  structured interconnected
    components, 45
  systems thinking, 47
  unit oriented
    architecture, 44, 46
Enterprise operations and
  customer interactions, 30
Enterprise prototype, 249
Enterprises
  abstract nature, 19–21
  accelerated digital timeline, 37
  architectural entities, 36
  erosive effects of time, 36
Entrepreneurial ecosystem, 365, 366
Evidence-Based Bootcamp, 359
Evolutionary Units
  building blocks of
    sociotechnical systems, 106
  collective formation, 108
  collectives, 111
  continuous evolution and
    functionality, 106
  developmental stages, 113
  fundamental building blocks, 88
  human civilization, 107, 113
  integrity and effectiveness, 109
  personal capabilities, 111
  social core, 107
  stability and integrity, 107
  stages, 110, 111
  survival and adaptation, 107
  technological extensions, 109
  technologically enhanced
    units, 113

INDEX

Evolution by organization, 92
Evolution by technological extension, 93, 94
Evolveburgh, 348–355
 education in, 357–361
 electricity grid, 352
 enterprise architects, 374–378
 politics, 369–373
 relativity square, 351
Excessive standardization, 33
Exchange of information, matter, and energy, 50
Exchange relations, 189
Existence (why), 137
Existential risks, 343, 344
Extended Star, 184, 185, 208
External engagement, 80–82

F

Fallacy of composition, 5
Feedback loops, 49, 96
Financial markets, 222
Financial Services Network, 241
Flat Network Organization with System-of-Systems Relationships, 220
Fleet integrative unit, 229
Flexibility and balance, 177
Formation of Fleets and Flotillas (Double Nesting), 230
Formation of Fleets (Nesting), 229
Forward-thinking strategy, 1
Function constructs, 259

Fused composition, 197–199
Future-readiness, 247

G

Geographical organization
 departmental segmentation, 278
 environment-based segmentation, 278
 naming conventions, 279, 280
 project-/service-based structure, 278
Gibberish eclipse clarity, 356
Global enterprise with geographical divisional structure, 67–69
Global Reach, 221–223
Google Cloud Interconnect, 125
Google Cloud Platform (GCP), 285, 286
Gratitude, 347
Great Enterprise Mystery, 13–15, 17, 23

H

Harmonious Integration of Realms, 218
Healthcare Delivery
 smart city initiative, 239, 240
Heterogeneous approach, 176, 177
Hierarchy, 49
Holism, 48
Homogeneous selection, 175, 176

INDEX

Hull integration, 256, 257
Human agents, 294, 295
Human-centered
    communication, 122
Human centricity, 114
Human coordination, 314, 315
Human-oriented values and
    objectives, 114
Hybrid selection, 177, 178
Hybrid structures, 183

I

Identical communication channels
    and protocols, 29
Identity and Access Management
    (IAM) policies, 285
Imagining and reimagining digital
    enterprises, 19
Imigrants, illegal, 379
Implementation Imperatives,
    UOEA, 247
Implementation, LSS
    code organization, 271
    engineering, 274–276
    iterative and adaptive, 270
    organization, 272–276
Indirect coordination, 119, 120,
    313, 314, 337
    stigmergy, 329
    unit-level interactions, 329
Indirect coordination and
    communication, 34
Information and communication, 28

Information security architecture, 4
Innate inclination for
    representation, 132
Innovation, 95
Integrated coordination, 79, 80
Integrated platforms, 307
Integrated systems, 51, 52
Integration challenges, 177
Integrative units, 293
Integrity inquisition, 360, 367
Interactions
    system-level, 296
    unit-level, 297
Interactive model
    communication, 136
Inter-cloud network routing, 125
Interdependence, 49, 95
Inter-entity relations, 28, 30
Interlocked and Fused
    compositions, 209
Interlocked composition, 196,
    197, 199
Internal reorganizations, 228
International logistics and
    Shipping, 241, 242
Interpretation (which), 137
Interstellar ventures, 362
Intranet, 123
Investment relations, 190, 191
IoT interfaces, 306

J

Job markets, 222

INDEX

K

Key performance indicators (KPIs), 311

L

Labyrinthine complexity, 211
Large enterprises, 211
Large sociotechnical system (LSS), 214, 254, 312
   aforementioned criteria, 149
   asynchronous communication, 338, 339
   component selection, 175-178
   compositional patterns, 192, 194-198
   comprehension of the enterprise's structure, 230
   cooperative interactions, 336, 337
   coordination, 334, 335
   decentralized control, 264
   digital facets, 148
   dual functionality, 256
   external entities, 231
   framework, 334
   historical precedents, 150
   implementation, 269
   indirect coordination, 337
   interactions with entities in environment, 231
   maritime formation, 199, 200
   metaphors, 147-149
   multi-faceted relationships, 335
   operational periphery (operational ships), 230
   physical sociotechnical systems, 149
   prototype for analysis and modeling, 186-188
   regional market, 228
   selection of components, 223-225
   selection of compositional patterns, 199
   structural pattern, 180-183
   structural uniformity, 253
   structured communication protocols, 336, 337
   structuring, 229
   transformability, scalability, and flexibility, 236, 237
Large systems
   prototype, 67-69
Learning capabilities, 50
Legitimacy, 233
Leveraging technology, 312
Local Area Network (LAN), 123
Logical fallacy, 5

M

Machine learning techniques, 315
Market shifts and competitive moves, 227
Memory constructs, 261
Mergers and acquisitions, 226

Mesh structures, 182
Metaphor
    conventional architectural paradigms, 145
    three-phased approach, 145
Metaphorical alignment with fleets, 211
Metaphorical foothold, 142–144
Metaphorical Imagination and Formation, 146
Metaphor's Aptness and Making the Final Choice, 147
Methodological Maze, 359
Military-derived command-and-control doctrines, 30
Misalignment, 114
Mobile interfaces, 305
Modular systems, 51, 52
Monolithic enterprise data platforms, 44
Multicloud approaches, 291
    AWS implementation, 282, 283
    Azure implementation, 283–285
    complexities, 280
    GCP implementation, 285, 286
    resilience and flexibility, 280
Multicloud Connectivity strategy, 125, 126
Multif-faceted relationships, 335
Multinational banking corporation, 241
Multitenant structure, 284
Mythological Sciences School (MSS), 358

N

Naming conventions, 279, 280, 290
Night vision tools, 17
Nonlinearity, 50
Notifications, 314

O

Operational applications, 256, 257
Operational components, 185
Operational Exchange, 78
Operational units, 294
Optimized communication networks, 238, 239
Order constructs, 260
Organization
    decentralization, 272
    functional requirements, 272
    geographical, 277–280
    multicloud approaches, 280–286
Organizational and digital architectures, 14
Organizational relations, 79, 189, 233
Organization and self-organization, 48
Organization (Org Node), 285

P

Paradigm shift, 316
Paranormal Psychology Institute (PPI), 358
Partitioned composition, 195, 196, 209, 256, 290

INDEX

Peer Review Gauntlet, 359
Peripheral integrative
    components, 184
Physical capabilities of humans, 57
Physical systems within
    sociotechnical contexts, 57
Political representation, 133
Politics in Evolveburgh, 369–373
Practical/functional
    representation, 133
Predict outcomes, 136
Process constructs, 260
Progressive complexity, 95
Progressive growth, 95
Prohibitive processes, 33
Public and government
    agencies, 222
Public transportation, 352
Purpose and exchange
    relations, 232
Purpose constructs, 259
Purpose-driven constructs
    core principles, 262–266
    culture constructs, 261
    digital constructs, 258, 259
    function constructs, 259
    identity manifestation, 259
    memory, 261
    order components, 260
    process constructs, 260
    purpose components, 259
    storage, 261, 262
    structure constructs, 260
Purpose-driven design, 251

Purpose-Driven Sociotechnical
    Wholeness, 219
Purpose Implementation
    chain, 263

Q

Quantum mechanics, 142

R

Real-time communication, 307
Reduced comprehensibility, 176
Reductionist approach, 4, 5
Regional fleet within
    Armada, 202–204
Regional Sales Unit, 235
Regulatory changes, 228
Relational patterns, 208
Relational patterns, LSS, 189–192
Remote Procedure Calls (RPC), 122
Reporting interfaces, 311
Resilience, 95
    and adaptability, 176
    of sociotechnical systems, 96
Retro-Futurism College (RFC), 357
Retro-Revolution Roundabout, 351
Ring structure, 181
Root Management Groups, 284

S

Scalability and flexibility, 212
Scientific inquiry, 359, 360

INDEX

Scientific representation, 133
Self-optimization, 317
Self-service interfaces, 300–302
Semiotic representation, 133
Service interfaces, 302–304
Seven Wonders of the Ancient World, 64
Single tenant structure, 284
Singularized modeling (unitary), 138
Smart city initiative, 239, 240
Smart Manufacturing Complex, 242, 243
Social capabilities (ability to relate), 60–62
Social representation, 133
Social system, 213
Sociotechnical Focus on Social and Digital Realms, 218
Sociotechnical systems, 43, 53, 54, 60, 88, 214, 293, 343, 345
   architecture, 69
   balanced integration, 65
   challenges, 106
   climate change and resource scarcity, 106
   connectivity solutions, 123
   coordination within, 118
   design of, 63
   economic growth and technological innovation, 105
   flexibility and scalability, 66
   foundational capabilities
   digital elements (the ability to compute), 63, 65
   natural or social resources, 63
   physical elements (the ability to make), 63, 65
   social elements (the ability to relate), 63, 65
   system's resources and benefits, 63
   fundamental architectural aspects, 69
   importance and impact, 104–106
   infrastructure, 104
   integration, 94, 118
   multifaceted interactions, 144
   scalability, 66
   social dynamics, 105
   sophisticated systems, 144
   sustainability and resilience, 106
Sociotechnical units, 251, 255, 262, 286
   accountability, 264
   autonomy, 265
   composition, 265
Software development, 4
Specialized optimization, 176
Stakeholder conflicts, 12
Standardization and modularization in ship design, 206
Star pattern, 182

INDEX

Stigmergy, 312
    BPM-executed coordination, 330–335
    coordination patterns, 315
    digital markers, 329
    disadvantages, 333–335
    embracing, 316
    intentional modifications, 330
Storage constructs, 261, 262
Strategic directions and boundary conditions, 233
Strategic placement of heterogeneous elements, 177
Structural integrity and control mechanisms within evolutionary units, 109
Structural layout of UOEA, 212
Structural uniformity, 267
    backing services, 255, 256
    building blocks, 254, 255
    hull integration, 256, 257
    partitioned composition, 256
Structure complex systems, 136
Structure constructs, 260
Structured business rules interfaces, 309, 310
Structured communication protocols, 336, 337
Supplier markets, 222
Synergy in Triadic Evolution, 95
Systemic integration, 79
System-level interactions, 296
    coordinating collaborative interactions, 312–318
    coordinating organizational interactions, 308–312
    coordination concept, 297, 298
    operational units, 297
    sociotechnical systems, 297
    transactional interactions, 299–309
Systems thinking, 47
System-to-system communication, 122

T

Technological—holistic development, balanced approach to evolution, 96
Technological periphery, 114
Technology evolution, 227
Technology stack, 270
Test hypotheses, 136
Traditional and modern coordination dynamics, 116
Traditional values, 354
Transactional interactions, 299, 300
    automated service interfaces, 304, 305
    challenges and solutions, 307, 308
    coordinating competitive interactions, 320–322
    coordinating compliance interactions, 318, 319
    coordinating cooperative interactions, 321, 322

coordination mechanisms, 306, 307
interface technologies, 305, 306
self-service interfaces, 300–302
service interfaces, 302–304
Transactional relations, 80
Transparency, 367
Tree, hierarchical structure, 182
Triadic evolution, 88
  biological evolution, 89
  components, 90–94
  enhanced systems thinking, 97, 99–102
  evolution by extension, 89
  evolution by organization, 89
  foundational principles, 103, 104
  history, 102
  holistic approach, 103
  human biology and organizational needs, 90
  human civilization, 90
  principles, 94–97
  social and technical factors, 103
  technological advancements, 90
Twelve-Factor App methodology, 255

U

Uniformity and consistency in architectural elements, 178–180
Unique customer requirements, 227
Unit, 216
Unit Architecture uniformity, 238
Unit-level interactions, 297
  direct coordination, 326–329
  indirect coordination, 329, 330
  individual units, 326
Unit orientation, 72, 75–81
Unit Oriented Architecture, 66
Unit Oriented Enterprise Architecture (UOEA), 68, 209, 212, 269, 344
  accountability, 246
  applications, 239
  benefits, 234, 244, 245
  built-in scalability and adaptability, 246
  drug development and clinical trials, 234
  functionality, 213
  human-driven insights and adaptability, 235
  integration, 246, 247
  integrative approach, 246
  principles, 217–220
  social and digital extensions, 246
  social system, 213
  sociotechnical system, 214
  strategic goals, 246
  system, 213
  technological resources, 235
  enhanced autonomy with strategic cohesion, 247

INDEX

Unit Oriented Enterprise
     Architecture (UOEA) (*cont.*)
  engineering, 274–276, 286–289
  organization, 272–276
Universal Application of UOEA
     Across Enterprise Types,
     243, 244

V

Vibrant community, 354–358
Virtual Private Cloud (VPC), 124

Virtual Private Network
     (VPN), 124
Virtual reality, 361

W, X, Y

Web interfaces, 305
Wide Area Network (WAN), 123

Z

Zachman, John, 41